THE COUNTRY WAY

THE COUNTRY WAY

By
Lloyd E. Eighme, PhD

*Cover Painting
and
Illustrations by*
Wayne Barber

Southern Publishing Association, Nashville, Tennessee

This book was
Edited by Gerald Wheeler
Designed by Dean Tucker
Text set 10/11 Palatino

Printed in U.S.A.

CONTENTS

A PHILOSOPHY OF LAND MANAGEMENT

"And the Lord God planted a garden eastward in Eden; and there he put the man whom he had formed" (Genesis 2:8).

What was that first garden like? If, as we believe, it was perfect in every way, it must indeed have been marvelous to behold. But God's plan did not intend that Adam and Eve should only sit on a mossy rock and watch perfect plants produce perfect flowers with perfect fruit. They were to tend the Garden. Man was to make the Garden of Eden a home for a family of human beings. He was not to let the plants grow at random but was to train and group them as functional parts of a lovely dwelling place.

God planned to have many garden homes established as the children of Adam and Eve spread outward from Eden. The Eden home would set the pattern for others. Satan thwarted the Lord by bringing sin into the first family. As a result, God had to bar Adam and his family from the beautiful Garden lest they continue to eat from the tree of life and become immortal sinners. But He had already instituted a plan of salvation which would someday reinstate humanity to the perfect garden. Until He could fulfill that plan, however, man must learn to live on an imperfect earth.

The generations since Adam have witnessed Satan's continued curse to our planet. The adversary has worked diligently to erase from man's mind the concept of life as God revealed it in the Garden of

Eden, and he has been all too successful. But the original design for man and his relation to the rest of creation remains unchanged. We can still realize some of the pleasure and satisfactions that were Adam's by making our homes as much like the Edenic one as possible in our present circumstances. The Lord will reward us, as we follow His instructions, with homes that are a little heaven on earth.

God always had some faithful believers, and He continued to enlighten them about how to avoid the pitfalls Satan had devised. God took many leaders of His people out of their artificial environments and brought them into closer contact with the natural world as a part of their training for leadership. Moses spent forty years in a pastoral setting before the Lord considered him prepared to direct the Hebrews in living the rural life-style of Canaan. His training in the courts of Egypt would have fully prepared him for a role in developing large centers of population with elaborate buildings. But God had something else in mind for the children of Israel. Their leader Moses would direct them back toward the pattern instituted in Eden. His life amid the hills of Midian helped to reorient his thinking in regard to the natural order of life.

Many years later the Lord spoke through the prophet Isaiah to remind His people that if they would be faithful and obedient to Him, they would receive the heritage of Jacob their ancestor. It was the same promise He gave to Abraham and Isaac to encourage them in complete obedience. Seventh-day Adventist Bible scholars have summarized the heritage of Jacob in the following statement found in the *SDA Bible Commentary,* Volume 4, pages 27, 28:

"God placed His people in Palestine, the crossroads of the ancient world, and provided them with every facility for becoming the greatest nation on the face of the earth (COL 288). It was His purpose to set them 'on high above all nations of the earth' (Deut. 28:1; PK 368, 369), with the result that 'all people of the earth' would recognize their superiority and call them 'blessed' (Mal. 3:10, 12). Unparalleled prosperity, both temporal and spiritual, was promised them as the reward for putting into practice the righteous and wise principles of heaven (Deut. 4:6-9; 7:12-15; 28:1-14; PK 368, 369, 701). It was to be the result of wholehearted cooperation with the will of God as revealed through the prophets, and of divine blessing added to human efforts (see DA 811, 827; cf. PP 211).

"The success of Israel was to be based on and to include:

"1. *Holiness of character* (Lev. 19:2; see on Matt. 5:48). Without this, the people of Israel would not qualify to receive the material blessings God designed to bestow upon them. Without this, the many advantages would only result in harm to themselves and to others. Their own characters were to be progressively ennobled and elevated, and to reflect more and more perfectly the attributes of the perfect character of God (Deut. 4:9; 28:1, 13, 14; 30:9, 10; see COL 288, 289). Spiritual prosperity was to prepare the way for material prosperity.

"2. *The blessings of health.* Feebleness and disease were to disappear entirely from Israel as the result of strict adherence to healthful principles (see Ex. 15:26; Deut. 7:13, 15; etc.; PP 378, 379; COL 288).

"3. *Superior intellect.* Cooperation with the natural laws of body and mind would result in ever-increasing mental strength, and the people of Israel would be blessed with vigor of intellect, keen discrimination, and sound judgment. They were to be far in advance of other nations in wisdom and understanding (PK 368). They were to become a nation of intellectual geniuses, and feebleness of mind would eventually have been unknown among them (see PP 378; cf. DA 827; COL 288).

"4. *Skill in agriculture and animal husbandry.* As the people cooperated with the directions God gave them in regard to the culture of the soil, the land would gradually be restored to Edenic fertility and beauty (Isa. 51:3). It would become an object lesson of the results of acting in harmony with moral, as with natural, law. Pests and diseases, flood and drought, crop failure—all these would eventually disappear. See Deut. 7:13; 28:2-8; Mal. 3:8-11; COL 289.

"5. *Superior craftsmanship.* The Hebrew people were to acquire wisdom and skill in all 'cunning work,' that is, a high degree of inventive genius and ability as artisans, for the manufacture of all kinds of utensils and mechanical devices. Technical know-how would render products 'made in Israel' superior to all others. See Ex. 31:2-6; 35:33, 35; COL 288.

"6. *Unparalleled prosperity.* 'Obedience to the law of God would make them marvels of prosperity before the nations of the world,' living witnesses to the greatness and majesty of God (Deut. 8:17, 18; 28:11-13; COL 288; DA 577).

"7. *National greatness.* As individuals and as a nation God proposed to furnish the people of Israel 'with every facility for becoming the greatest nation on the earth' (COL 288; see Deut. 4:6-8; 7:6, 14; 28:1; Jer. 33:9; Mal. 3:12; PP 273, 314; Ed 40; DA 577). He purposed to make them an honor to His name and a blessing to the nations about them (Ed 40; COL 286)."

Notice particularly the promise of skill in agriculture and animal husbandry. It almost staggers the imagination to contemplate its fulfillment today. Of course it was conditional to a particular group of people, but it illustrated the desire of the Creator to take His people back to the original plan of harmony between Adam (man) and the Garden of Eden (his environment). Can you grasp the significance of Israel's growing crops with no pests or diseases, no floods or droughts? Such a thing is hard to visualize in our pest-ridden, unbalanced ecosystem of today.

How much of the promise did God fulfill? The record is brief, but we can read in Scripture about Israel's becoming established in Palestine. Their obedience was rather short-lived, and it is difficult to ascertain to what extent they realized the completion of such promises. I have often wished for a written account of the methods in agriculture and animal husbandry that the Lord planned to teach to the Hebrews. A careful search of the Scriptures will reveal a few, but only a few, specific concepts. Why didn't Moses include more details? We can only speculate that, if given all the information at the beginning, the people might have used it for their own selfish gratification instead of for their Creator's glory. The fact that Israel failed to obey God's principal commands would explain their failure to receive the knowledge that would have restored Canaan to its Edenic state.

Modern history records the sad condition of the area after Israel's rejection of the Messiah. The great cedars of Lebanon once flourished over a forested area of nearly two thousand square miles. King Solomon made an agreement with Hiram, King of Tyre, to furnish cedar for beams and panels to beautify the Temple. The extensive lumbering operation required eighty thousand lumberjacks and seventy thousand men to move the logs to the sea. Today the cedars of Lebanon are practically extinct. Only a few small groves struggle for survival. A church protected one grove, making it sacred. A stone fence around the church kept the goats out. Within the fenced area seedling cedars have sprung up into a dense forest of straight, fast-growing trees.

Overgrazing has destroyed the forests that would have protected valuable topsoil from washing into the valleys. The rich soil that once supported great flocks and herds from the days of Abraham and produced food for the great nations of antiquity has vanished,

leaving only rocks and thistles that will not support
even the wild goats.

The fine topsoils from the headwaters of the Tigris
and Euphrates rivers have formed a muddy delta ex-
tending eighty miles into the Persian Gulf. The modern Israeli people have worked diligently to rebuild
some of the devastated land and have achieved remarkable results with modern agricultural techniques. They will never be able, however, to put the
rich topsoil back on the hillsides or reestablish the
original vegetation.

What did ancient Israel do wrong that led to such
complete devastation of the land of Canaan? First of
all, by their failures to obey the Lord they forfeited the
promised blessing on the land. The Book of Isaiah
pronounced many disasters upon the unfaithful and
disobedient people. It predicted drought and famine
and condemned certain life patterns. In Isaiah 5:8 the
prophet declares, "Woe unto them that join house to
house, that lay field to field, till there be no place, that
they may be placed alone in the midst of the earth!"
All families in Israel received a piece of land as an
inheritance that no one could take from them. Even if
they chose to sell or lease their land to someone else, it
would revert back to them at the end of an established
number of years. Everyone at least had the opportunity to live on his own land. If the people had followed
such a plan, they would have had no large concentrations of population in cities. Even Jerusalem was
planned as a city of worship where the people could
gather together for religious festivals and sacred
ceremonies, not as a metropolis dependent on the
adjacent farmland to supply it with food. The land
would produce in great abundance, but without a vast
system of transportation, the demand would quickly
exhaust and wear out the fields surrounding a size-
able city.

One of the few admonitions we do find on how

God wanted His people to manage the soil involves a seven-year plan calling for crop rotation and a full year of rest for the soil. That would allow the beneficial bacteria and other organisms a chance to replenish the soil with nutrients. Such a scheme worked in a rural setting where each family owned sufficient land. Near large cities, however, they would have to continually use the land to support the urban masses who could not grow their own food. It seems obvious that the Lord's plan never included people crowded together. The ruins of ancient cities not only reflect how the people failed to receive God's promised blessing but also how they destroyed the soil around them.

In the world today we still find the greatest problems of mankind in the cities. As the industrial revolution swept across the American continent, we saw large cities and centers of industry develop which became directly dependent upon the immediate area for support. The 1937 documentary film "The River" graphically portrays it: "We built a hundred cities and a thousand towns, but at what a cost!" We slaughtered our forests from the Atlantic Ocean to the Great Lakes and south to the Gulf of Mexico to provide lumber and fuel for the large cities. The land we wore out by growing cotton and tobacco. The prairies we plowed up to grow wheat for the cities and saw our precious topsoil blown and washed into the Gulf of Mexico. To build the city of San Francisco twice, we slashed and burned our primeval redwood forest, wasting as much in the logging and milling operations as the earthquake and fire destroyed. The increasing demands for lumber, plywood, and pulpwood to build our expanding megalopolises have decimated the great fir forests of the Pacific Northwest that once seemed so inexhaustible.

Is it impossible to turn the tide of people from the cities back to the country now? Is our exploding population so far out of hand that we no longer have enough land to support man on the earth? Those who answer "Yes" to such questions, I believe, turn their backs on God's promised blessings. Even though the Lord offered the heritage of Jacob to a specific nation at a particular time, the principles of the promise can apply to all who become a part of spiritual Israel by submitting in faithful obedience to the same God who still has the power to fulfill them in our lives. We must show our willingness to follow the instructions we already have before we can expect greater blessings. If it is not God's plan for man to be crowded into large cities, then we should do everything possible to get out of the cities and back to the land.

Some will say, "It is impossible to move everyone out of the cities; many would not even be able to survive." That is undoubtedly true, in addition to the fact that probably we could convince only a few to leave anyway. But it does not prevent those who are ready and able to adjust to the country from doing so. Furthermore, the food such country dwellers produce, even those who remain near enough to commute to a job in the city, becomes a real factor in supplying food for the world without damaging our precious heritage, the soil. In Russia, for example, the food raised in the backyard gardens of the commune laborers is a more reliable source than that of the farms themselves.

Vacating the city for the country is not something to do on impulse. It demands a great amount of careful planning and education to avoid financial disaster and unhappiness. Some may find suburban living to be a temporary solution. The information in this book should be of value both to the suburbanite as well as the truly rural family. Country living should be a way of life adaptable to a variety of circumstances.

CHAPTER 1

MODERN LAND MANAGEMENT PATTERNS

Complexity in the ecosystem tends toward stability. Or, to put it another way, the more different kinds of plants and animals that live together, the more likely each kind will thrive.

That is a basic ecological principle, but how does it actually work in practice? Let's take a common food crop as an illustration. Potatoes—our so-called Irish potato, taken from its native Central and South American highlands to Europe by the early Spanish explorers—offer a good case in point. After some improvements by plant breeders, it became the main food crop in Ireland. The northern climate was particularly favorable for potatoes, and the people soon learned that a good cellarful of them could provide the main source of nutrition for the long cold winter months. In fact, they became so much a part of the Irish diet that the people began to grow them to the exclusion of almost all other vegetables. The countryside became essentially one large potato field.

About 1845 the late-blight disease of the potato spread rapidly throughout Ireland, and in two years it almost wiped out the entire crop. Starvation and misery forced many people to flee to other countries such as America. If your ancestors came from Ireland in about 1846, you may be an American today because of the late-blight disease.

Why did the disease suddenly wreak such havoc in Ireland and not elsewhere in the world? Was that the only place where potatoes grew? First, consider the fact that the Irish farmers grew practically nothing but potatoes; they had eliminated almost all other crops. With such a large continuous planting of one kind of plant, any pest or disease organism would find ideal conditions for multiplying and spreading. With no barriers of other vegetation to separate one potato field from another, the disease could spread easily across the country. Late blight of potatoes, a fungus disease transmitted by microscopic spores, infects the growing stems and leaves. It also spreads to the tubers of the plants that survive its attack. The farmers used the infested tubers as seed for the next crop. It would have taken several years for the disease to expand through a more varied and complex ecosystem, thus giving the farmers a chance to switch to new or resistant crops.

The late-blight disease took its toll of potato crops in other lands, also, including America. However, elsewhere the pattern of agriculture was more varied and potato fields were more scattered. Farmers there discovered ways to continue growing the vegetable in spite of the disease. In the United States, however, another monoculture problem appeared also in connection with potatoes.

Across the great plains to Colorado dwelled a native weed called the buffalo bur. It was a member of the nightshade family, along with tomatoes, deadly nightshade, tobacco, and potatoes. The buffalo bur weed did not occur in great expanses but existed scattered through the complex ecosystem of many species of weeds, grasses, trees, and shrubs. A seemingly harmless round, tan, striped beetle fed on the buffalo bur. The beetle population could not multiply beyond its food supply because it had to compete with many other insects in the complex ecosystem, and it spent considerable time and energy just moving from one weed to another. As large cities developed, a ready market for potatoes led farmers to plant larger and larger acreages to sell for cash income. When the beetle found that potato leaves tasted just as good as buffalo bur, it discovered a food supply such as it had never had before. Food is usually the greatest limiting factor in the growth of any population. Just supply sufficient food, and you will see how fast living things can increase.

For example, consider the little fruit fly, the tiny gnats that gather around ripening fruit in late summer. If you placed one pair under *ideal* conditions with an unlimited food supply, they would mate and the female would lay about 100 eggs, from which 50 males and 50 females would hatch. Upon reaching maturity each of the 50 females could produce another 100 eggs. The life cycle of the fruit fly is so short that it can proceed through 25 generations in one year. If the flies bred under such conditions through 25 generations, and we packed only the flies of the twenty-fifth generation into a solid ball, it would be 93 million miles in diameter, reaching from the earth to the sun. Just start multiplying 50 by itself a few times and see what kind of numbers you get. This is biotic potential unleashed. Why doesn't the present world quickly fill up with fruit flies? Mainly because only a few of each generation find enough food to survive and reproduce.

When the striped beetle found a new food supply that allowed it to more fully express its reproductive potential, it became known as the Colorado potato beetle, a serious detriment to growers who planted large acreages. The parasites and predators that had once helped to limit the beetle, mostly stayed behind in the complex ecosystem of the buffalo bur weed. They were not capable of moving into the vast monoculture of potato plants. Even the birds that had used the beetles for food will not likely follow them out into a forty-acre potato field. The birds prefer the varied ecosystem of hedgerows and natural vegetation. With most of its enemies left behind and a vast supply of suitable food before it, is it any wonder that the Colorado potato beetle overwhelmed large tracts?

I remember listening to my father tell about the unpleasant assignment he had as a youngster walking up and down the long rows of potatoes, picking off the pink, slippery larvae or grubs of the beetles and destroying them—an almost impossible task because of their rapid-breeding rate. Farmers used Paris green and other poisons. But even DDT, chlordane, Malathion, and all the other insecticides have not eliminated the pest. We have tamed the atom and landed on the moon, but we are still losing the battle in the war between insects and man.

Can't we push the Colorado potato beetle back to the buffalo bur that it once found satisfactory? Not as long as farming, with its pattern of widespread single crops, offers better and bigger food sources. If the ecological principle that complexity tends toward stability is really true, can we stabilize the potato-beetle population from exploding by diversifying the ecosystem?

Yes, science has stated in a number of ways that the interacting of all of the members of a complex ecosystem hold each other to stable population sizes. Such a situation would occur if we planted potatoes with other vegetables and native plants. Don't forget that Colorado potato beetles will exist wherever potato plants grow unless we completely eliminate them from that ecosystem. Increased variety in the complex ecosystem, however, will cause it to approach the original pattern in which the insects lived for centuries on the buffalo bur without ever destroying the plant.

Is it possible to eliminate the present pattern of large-scale single crops? Probably not. At least not as long as we depend upon commercial growers for the bulk of our food supply. They have developed a marvelously efficient system in an attempt to earn a living despite low crop prices and constantly increasing labor costs. Such agriculture makes the United States one of the greatest food-raising nations of the world. As long as we demand the lowest possible food prices and fruits and vegetables with no blemishes from insect damage, we continue to promote the vast monoculture and its greatly simplified ecology. Modern technology has produced artificial controls for the pests and diseases that have threatened our unstable monoculture during the past fifty years. How long the pattern of barely outrunning the pest threat can continue we can only speculate, but no end to it appears in the near future.

The next logical question seems to be: Is there any way to reverse the trend from unstable monoculture to more stable, varied ones? That would mean a return to the kind of ecosystem existing in the days of our pioneers. It was more complex and therefore less subject to total destruction. We should not imply that the early settlers never struggled with pests and diseases in their gardens. Requests for governmental help and advice from small farmers and home gardeners led to the forming of our state and federal departments of agriculture during the nineteenth century. The pamphlets and bulletins issued in those early years make interesting reading.

A publication from the year 1797 gives advice on how to control the cankerworm, a caterpillar that chews the leaves from fruit trees: ". . . burning brimstone under the trees in a calm time;—or piling dry ashes, or dry, loose sand, round the root of trees in the spring;—or throwing powdered quicklime, or soot, over the trees when they are wet;—or sprinkling them, about the beginning of June, with sea water, or water in which wormwood or walnut leaves have been boiled;—or with an infusion of elder. . . . The liquid may be safely applied to all the parts of a tree by a large wooden syringe, or squirt."

The remedies suggested were usually more cultural than chemical, and we wonder if they really worked. No modern farmer today would consider such methods as suitable. Such approaches were geared to a much smaller scale of farming and for a different style of consumer.

Prior to 1900 people usually took for granted the fact that some of the apples would be wormy and vegetables would show signs of insect damage. Today's modern housewife who discovers that one of the shiny red apples in the plastic bag formerly served as the home of a worm is ready to bring a lawsuit against the food chain store that sold her defective merchandise. When the consumers picked their own apples from their own trees, they had no one to blame for wormy apples other than themselves or Mother Nature. People in those days were skilled in using a knife to excise the wormy portion of an apple, and sharing an apple with a worm was not upsetting unless they discovered only half of the worm.

It really is not necessary to give in to the insect competitors for our food supply and accept defeat. We can move away from unstable ecosystems toward more stable ones by reeducating the consumer on what he should expect in prices and food appearance.

One of the first steps to take in this new direction is to encourage more people to grow their own food. If everyone with a suitable spot for a garden would make good use of it and grow as much of his own food as possible, it would reduce the demand from the large producers. It may seem like a small thing, but at least it would be a step in the right direction. Food grown in the backyard garden would form part of a more complex ecosystem, one less subject to large outbreaks of pests. Where insects do pose a problem, one could control them by simple practices that work well on a small scale but are not feasible in a large monoculture system. In most cases the labor involved in small garden projects would not add to the cost of the food but rather provide healthful outdoor exercise. One could make a whole list of benefits from growing one's own food, not the least of which would be fresher, better-flavored, more-nutritious, and less-expensive produce.

Small gardens with a mixture of plants including flowers and native shrubs, trees, and weeds usually support a wide variety of parasites and predators that check the growth of insect populations before they get out of hand. On several occasions I have seen such natural checks and balances working on my own little acre. In the spring of the year the apple trees make rapid tender-leaf growth particularly attractive to aphids. If not controlled, the aphids cause the leaves to curl into a sticky mess and deform and stunt the little green apples. I have watched anxiously as the aphid population builds upon my trees in spite of my attempts to disrupt the busy ants that carry them from twig to twig. Native vegetation, brush, and trees surround my garden, and the garden itself is a mixture of fruit trees, berries, vegetables, and flowers.

As the aphid population becomes established, I see ladybugs, lacewings, snake flies, and parasitic and predatory wasps begin to move in from the surrounding vegetation. Most years the damage from the aphids is slight, and I remove the few stunted apples at thinning time. In poor years, however, the aphids reduce the yield of my trees to the point where, if I depended on them for a living, I would be in trouble. But even with aphid damage we usually have more apples than we can use. Again we see the difference between farming for a living or merely growing your own food. If I used my apples and other fruit for an income, I would have to do everything possible to produce as large a crop as possible every year.

Some agroecosystems—a fancy name for all the

living things in a garden or crop field—are inter-
spersed with natural vegetation, and they thereby
benefit from the competition among insect populations. In the Napa Valley of northern California,
famous for its wines, a tiny leafhopper attacks the
grapevines. It has the potential to do a great deal of
damage but seldom does. In the San Joaquin Valley, a
few miles away, the same leafhopper is a serious
problem, held in check only by spraying and dusting.
Biologists have discovered why the leafhopper does
not present as great a hazard in the Napa Valley. They
studied the parasites and predators of the leafhopper
and found a tiny wasp parasite to be the most effective
biological control agent. The parasite cannot over-
winter on the leafless grapevines, but it spends the
winter months in blackberry brambles along stream
banks. When spring comes and the leafhopper
population increases, the wasps move from the
blackberry brambles into the vineyards and parasitize
the leafhoppers.

The Napa Valley is long and narrow and inter-
spersed with streamside vegetation which includes
blackberry brambles and wild grapevines. No vine-
yard is far enough from native vegetation to permit
the leafhoppers to multiply unhindered by the parasites. The San Joaquin Valley, on the other hand, is
many miles across and lacks the streamside vegetation so essential for the overwintering wasps. The
leafhopper populations have no natural checks there,
and farmers must resort to artificial controls. Before
man ever brought European grapes to the Napa Valley, a balanced population of grape leafhoppers and
their parasites lived in the natural streamside vegetation with its wild grapes and other native plants.

Grape phylloxera
grafting on american roots

CHAPTER 2

HOW TO MAINTAIN SOIL FERTILITY

Many people think of soil as just "dirt." It may be part of the reason why they neglect and misuse it. Understanding how the soil is a living system that functions properly only when healthy is essential to maintaining soil fertility. Likewise, an understanding of how the human body functions is vital to human health. Soil can grow and improve, or it can die and be destroyed. The living soil is a microecosystem and requires careful management like any other part of the total system. Any practice that damages the soil will result in poor crops and poor people.

Soil consists of three basic components:

1. Minerals such as rocks, sand, and clay
2. Decaying organic matter
3. Living organisms

The mineral component is what most people think of first. It slowly forms through the erosion of rock into smaller stones, sand, and clay. A continual process, we can observe it around glaciers and rivers. The beautiful blue color of some mountain lakes such as Lake Louise in the Canadian Rockies, or others fed by water from large glaciers, results from the extremely fine rock flour suspended in the water. The constant grinding as the ice moves the loose surface boulders against solid rock creates the dust. Glacial rivers are often milky with rock flour and fine particles of mineral—a fact we learned once on a camping trip when we dipped water for the soup kettle from a river in the rugged Cascade Mountains of Washington

State. The rock flour settled between our teeth and under our tongues like so much scouring powder. If we had let the water bucket set for even five minutes before pouring it into the kettle, the heavier mineral particles would have drifted to the bottom and the finer rock flour would have remained as an unnoticed addition to the soup. It is the latter which gives the beautiful opaque color to glacial lakes. When it does eventually settle out, it forms a layer of sticky clay. Weathering and mixing of clay and larger particles with organic matter eventually produces the loamy soil that we look for in our garden.

Productive soil is more than just a mixture of sand and clay. It must contain nutrients to support plant growth. Nitrogen is the main one, and it must be present in a form that plants can use. A marvelous recycling system continuously replenishes the soil with nitrogen. As plants grow they use the minerals in the soil together with carbon, hydrogen, and oxygen from the air along with water to build leaf and stem structures. The plant completes its life cycle, and its leaves and stems drop to the soil. Organic matter, the second of the three soil components, recycles into nutrients again through the activity of living things in the soil.

The most important soil organisms are bacteria. Recycling soil nutrients requires a large and diverse group of bacteria known as decomposers. But they need water, warmth, food, and air. No form of life can exist without water, as the lunar astronauts have observed on the arid moon. Bacteria must have moisture in which to live, feed, and reproduce.

I will never forget my first trip across the dry California and Arizona desert. The litter and debris scattered from the highway out into the desert perplexed me. At first I assumed that we were following the route of the garbage disposal trucks from a large city. After many miles with no improvement, I concluded that I was seeing the accumulation of many months and years of material that would, in climates with more moisture, soon break down and return to the soil. With no moisture available, the decomposing bacteria could not do their job. Even the bleached remains of dead plants among the sparse desert vegetation appeared almost indestructible. A few dried cactus stems showed evidence of termites working from within where they could conserve every molecule of water. But the desert soil has an extremely slow recycling program and cannot support much plant growth.

Bacteria active as decomposers require warmth. Since they cannot maintain body heat, winter cold brings them to a complete standstill. Have you ever visited a popular ski area just as the last of the snow melts? The candy wrappers and other paper debris dropped by the retreating snow and the entire winter accumulation clutter the ground for a short time before the warm sun and abundant moisture aid the bacteria in their task.

All living organisms must have food with a balance of nutrients to support life and growth. Bacteria are no exception. If you supply them with only clean straw or fresh sawdust, they will starve to death. They must have some form of nitrogen.

Most bacteria must also have air to breathe, though a few exceptions do exist. If organic material becomes so wet and soggy that air can't get in, only those few kinds of bacteria that can live without air will survive. The mass will turn putrid and sticky instead of breaking down into the spongy portion of the soil that we refer to as decaying organic matter.

One large group of bacteria have the special ability to take nitrogen gas from the air and convert it into a form which plants and other bacteria can use. Such nitrogen-fixing bacteria live in root nodules on legumes such as clovers and peas. Of all the different kinds of plants and animals that dwell in the soil, only they can utilize the nitrogen gas that makes up 80 percent of the atmosphere.

We should not neglect to mention here other organisms essential to good soil. Protozoa, fungi, mites, and insects provide a basic food supply for larger soil organisms. Earthworms, great cultivators of the soil, work with the bacteria in breaking down organic matter.

Plants must have a balanced diet in order to grow properly. The Edenic soil was no doubt perfect in texture and composition so that any plant could grow to its full development and reproduce more of its kind. The curse of sin has let the soil degenerate, and it is only by careful management that we can maintain its fertility.

In nature's plan, plants are the great producers and animals are the consumers. Through photosynthesis, green plants absorb the sun's energy and convert it into food for animals. Without green plants, no animals could survive on earth. The raw materials for photosynthesis are carbon dioxide and water. The air contains a plentiful supply of carbon dioxide, and the plant obtains water from the soil. But to maintain photosynthesis, the plant must have roots to take up water and stems to support the leaves that capture the energy of sunlight. Also it must have complex chemical compounds like chlorophyll and the various enzymes that promote the transfer of energy from the sun to the food that the plant manufactures.

Three basic nutrients that the growing plant requires in large quantities are nitrogen, phosphorus, and potassium. We often refer to them by their chemical symbols N, P, and K, the nitrate, phosphate, and potash of fertilizers. Nitrogen produces lush, green growth, and a lack of it usually results in stunted, yellowish vegetation. The chemical is an essential part of proteins which plants manufacture to build the

PHOTOSYNTHESIS

WATER + CARBON DIOXIDE →(SUNLIGHT / CHLOROPHYLL)→ CARBOHYDRATE + OXYGEN

protoplasm within their cells. Phosphorus keeps photosynthesis going. The plant will be stunted if it cannot get enough phosphorus, and it will stop growing. Potassium is vital for the synthesis of food materials stored up by the plant. Without potassium, the plant will not form strong stems and will fail to produce seeds and fruits.

Poor soil may lack six other important chemical elements: calcium, sulfur, magnesium, iron, zinc, and manganese. Growing cells require calcium, and sulfur is an important ingredient of cell protoplasm. The center of every chlorophyll molecule contains magnesium. Iron, zinc, and manganese we often label as trace elements since the plant employs such minute quantities. But without them a plant cannot function properly. Yellowing leaves indicate iron deficiency.

Of all of them, the one most difficult to maintain in sufficient quantities in the soil is probably nitrogen. Nitrogen, like other nutrients, passes into the plant from the soil through the tiny root hairs which can absorb it only in a water-soluble form. Rain and irrigation, however, quickly wash such a solution away; therefore we must continually replace it to ensure an adequate supply.

Most gardeners ask at some time or other about the best way to provide a continuous and plentiful supply of nutrients. To answer the question we must first understand how the plants extract nutrients from the soil.

Plant roots do not have mouthlike structures that can open to ingest food particles. The root hairs must absorb everything. If you have never seen root hairs, observe them on the newly developing root from a seed. Radish seeds sprout quickly. Should you have some left over from last year's garden, put a few on a damp piece of paper towel and slip it into a plastic bag to keep it from drying out. At normal room temperature the seeds will sprout in a day or two. In

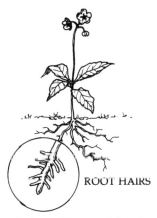

ROOT HAIRS

moist air the root hairs will resemble a band of white mold just back of the root tip. Looking at the root under a magnifying glass, you will observe that the root hairs are slender threadlike extensions of the surface cells. A spot not larger than a pinhead may have as many as one hundred root hairs. Take the sprouted seeds out of their moist environment, however, and the root hairs will quickly dry up and shrivel away.

Root hairs are tubular like the fingers of a glove. The nutrients pass through the living-cell membranes that form the wall of the tube. Even a microscope will fail to reveal any openings though. Whatever goes through the membrane must be in the form of chemical molecules dissolved in water, and even then only chemicals of small molecular size. The process is similar to the human digestive system. The microscopic villi, or fingerlike projections that line the intestine, can absorb only small molecules. Digestion breaks the large food molecules into substances like glucose, amino acids, and fatty acids which are minute enough to enter the bloodstream in a water solution. Likewise, only small chemical molecules in solution can penetrate the root-hair membranes.

When we supply a plant with nutrients in the form of organic matter, the decomposing bacteria do the "digesting" or breaking down into chemical molecules suitable for absorption. If the organic matter contains all of the essential elements in proper proportions, then all of the chemical substances necessary for growth will enter through the root hairs as the bacteria release them. But if both decomposed organic matter and chemical fertilizers supply the same nutrients, then what advantage do organic fertilizers have—if any? Several. First of all, if the organic matter comes from decaying plants, then all of the nutrients incorporated in that plant material should become available to the new plant. Plant material grown in soil containing all of the required chemical substances should provide a balanced plant diet. Secondly, organic matter will release its nutrients slowly during the time the plant needs them. Another advantage involves the improved texture of the soil that results from adding organic matter.

Will all organic matter provide a balanced diet for growing plants? Not necessarily. Especially not if the gardener uses only one kind exclusively. For instance, plant material grown in soil deficient in some nutrients cannot provide substances unavailable to it. However, if plants will survive at all in the soil, you can be sure that it contains everything absolutely vital to growth. Some elements may be in short supply, though, and produce stunted or unhealthy plants. They would not be as good a nutrient source as plants raised in balanced soil. If you employ only one part of the plants, such as sawdust or straw, you may create a lack of nutrients normally found in the other areas of the plant.

How can you determine if the soil has a good supply of nutrients? Let the plants tell you. They can communicate their needs to you by their appearance if you learn to read certain signs. A lack of nitrogen shows up in pale, stunted growth. Insufficient phosphorus will also produce stunting, but the leaves will be dark green instead of pale. A deficiency of potassium prevents the plant from storing up food in the fruits or roots.

It is possible to overfeed a plant. Even an overabundance of trace elements can harm it. Since nitrogen promotes green growth, too much of it will result in excessive leafy foliage. I have seen tomato plants that produced so much foliage that the stems would break, and they continued to grow leaves instead of setting fruit. Too much of any nutrient can form such concentrated solutions that it destroys the root hairs. Then the plant will die even with ample food in the soil. If the chemical concentration in the soil water is too great, it will draw water out of the roots, "burning" the plant—an obvious sign of excessive feeding. The plant may wilt easily, and the edges of the leaves will turn dry and brown. Wash the fertilizer out of the soil with large amounts of water, or the plant will continue to dry up and die.

Overfeeding of plant nutrients can occur even with organic fertilizers. Undecayed animal manures, especially chicken manure high in nitrates, can produce the same damage as synthetic chemical fertilizers. All animal manures, especially those that have retained large amounts of the urine or liquid wastes, contain concentrations of chemicals such as salt that will eventually build up in the soil and interfere with normal growth.

A plant can't discriminate between organic and synthetic chemical fertilizers. Why then do so many people shun synthetic fertilizers and turn to organic ones? Remember the numbers on the package of snythetic fertilizers such as 10-6-4 or 16-20-0 that indicate the proportion of NPK, or nitrogen, phosphorus, and potassium. Those numbers represent percentages of available nutrients. A 100-pound bag of

10-6-4 fertilizer contains 10 pounds of nitrogen, 6 pounds of phosphorus, and 4 pounds of potassium in some form that the plant can absorb. That accounts for 20 pounds of the 100-pound bag, so what is the remaining 80 pounds? It is usually sand or some other inert material.

Why don't the fertilizer manufacturers sell you the pure plant nutrients instead of a dilute mixture with so much inert material? For one reason, it would cost too much to carry out a purification process to that extent. Another more important reason is that you would find it difficult to apply the pure chemicals in small enough amounts to avoid burning your plants. In other words, it may be less expensive for you to pay the shipping costs of 80 pounds of inert material to get 20 pounds of nutrients than the expense of completely refining the chemical nutrients. Second, it gives you a product much safer to use. Here we find the answer to our earlier question about the hazards of synthetic chemical fertilizers. The NPK formula for cow manure is 1-1-1. A 10-6-4 fertilizer, even though it is relatively low in nutrients compared to more concentrated ones, is much more likely to "burn" your plants than will cow manure—the main reason why many people prefer organic fertilizers.

The old adage "If a little is good, a lot is better" has led too many people to apply increasingly larger doses of fertilizers to their soil in an effort to increase production. But in doing so they have overfed their plants to the point of "poisoning" them with high concentrations of chemical nutrients. It is so similar to some problems of human nutrition that I cannot resist the opportunity to draw a parallel. We talk about the great damage done to the human body by overeating concentrated, highly refined foods such as candy and pure cane sugar. The evidence is so convincing that many have attempted to leave all sugar out of their diet. What they too often don't realize is that sugar, in a certain form called glucose, is one of the three basic nutrients absorbed into our bloodstream. Whether you get it in a nonrefined form such as fruit or a refined chemical from sugar beets, the body will absorb it into your bloodstream as glucose. Therefore the real problem is not so much which kind of sugar you eat, but rather how much of it you consume. Refined foods are to humans like high-formula synthetic chemical fertilizers are to plants. The valuable chemical nutrients are there, but in such a concentrated form that you must apply them carefully to avoid upsetting the chemical balance in the plant's nutrition. Unless you really understand what you are doing, you should work with less concentrated forms of fertilizer.

Likewise, we can compare the "empty calorie" foods in the human diet to the simplified nutrients in some synthetic chemical fertilizers. The so-called "organic" source of plant nutrients, such as animal manures or compost, if properly used, is more likely to contain the trace elements and other substances vital to nutrition than a simple chemical compound like ammonium nitrate which supplies only nitrogen. However, let us not lose sight of the fact that the nitrogen from pure ammonium nitrate is chemically the same as that from cow manure, and the plant can employ it as part of a balanced diet.

Another hazard of chemical fertilizers concerns soil management more than plant nutrition. Any experienced gardener recognizes the advantages of soft, loose soil rich in organic matter over hard, compacted soil with little or no organic material. If year after year he takes a crop and adds nothing other than the exact chemical nutrients needed to grow that crop, the soil will eventually become hard and difficult to manage. Without sufficient organic matter in the soil the plants can't develop healthy root systems. Organic matter acts as a sponge to hold moisture and nutrients and helps air and rain to penetrate the soil. It also serves as a buffer to keep the soil from going extremely acid or alkaline. Such benefits are sometimes more important than the nutrients themselves in organic matter because, without a healthy soil environment, the plant cannot survive long enough to use the supplied food substances. Employing only synthetic chemical fertilizers often leads to a lack of organic matter and poor soil structure.

What do we mean by organic fertilizers and organic matter? The word *organic* simply refers to the carbon-containing chemical compounds produced by living organisms. When we say organic matter, we usually mean the bulky remains of dead plants and animals or plant and animal products such as leaves and manures. When a farmer plows under a green cover crop, disks the stubble into the soil, or spreads animal manure on his fields, he is an organic farmer. If he applies synthetic chemical fertilizers to his fields, he gets classified as a nonorganic farmer. Could he be both at the same time? Obviously he could.

Some apply the terms "artificial" or "unnatural" to nonorganic fertilizers. I don't like the connotation. We do not originate the chemical substances in a package of fertilizer. The elements of nitrogen, phosphorus, and potassium God brought into existence at the time of Creation. We may sort them out into a concentrated form, but they are the same elements found in the living cells of a plant.

Is there any final conclusion then as to which method of gardening is best, organic or nonorganic? If we define nonorganic methods as using no organic matter, then we must practically restrict nonorganic gardening to hydroponics or soilless culture. Even soilless culture we could consider organic in the sense that some of the chemicals employed may be organic compounds. I think most people would not carry the distinction that far but would more likely refer to nonorganic gardeners as those who use chemicals. Yet the chemical nature of all plant nutrients as absorbed by the plant invalidates the argument that organic gardeners do not use chemicals.

We must be concerned with the form in which we supply plant nutrients. An organic gardener would attempt to supply all nutrients in some form that we would call organic matter, even including concentrated forms such as dried blood and hoof-and-horn meal. The nonorganic gardener would most likely turn to a combination of organic matter and synthetic chemical fertilizer. (At this point we are thinking of plant nutrients only and not the chemicals involved in pest control.)

After considering the hazards of synthetic chemical fertilizers, it would seem that the safest path for a beginning gardener would be the organic way. Obviously the damage to plants and soil by synthetic chemical fertilizers results from misuse. The proper handling of concentrated chemicals to supply a balanced diet for plants is difficult for amateurs. The temptation is too great to overfeed and neglect the important bulk so essential to good soil. It would seem that the ideal diet for plants would be as near as possible to the way God originally set it up. If such a thing as an undisturbed ecosystem exists anywhere on our planet today, it should teach us something about the original plan.

Oak trees and pine trees continue to thrive in the forest where man has not interfered. Where do they get their nutrients? A continuous recycling process takes place as leaves and twigs fall to the ground and the decomposing bacteria digest them into forms of nitrogen, phosphorus, and potassium that the trees can take in. Being nature's own compost heap, it works well. But somewhere we must alter the ecosystem to grow carrots and beans and potatoes. If our vegetable garden had exactly the same nutrient requirements as the oak and pine trees, we could just gather oak leaves and pine needles to make compost for the vegetable garden and have a balanced diet for our food plants. But we have discovered that the great range of plants we grow in our garden will do better on a more varied diet than just oak leaves and pine needles. So we gather kitchen scraps, animal manure, weeds, and lawn clippings, and make a compost heap that should supply a more complete set of nutrients for our garden.

What if we cannot find the right mixture of organic materials in sufficient quantities for our compost heap? Should we limp along with scanty, under-nourished plants, or should we risk using some chemical fertilizers? It is a decision that every individual must make for himself. Wherever feasible, balanced organic gardening methods should yield the best results over a long period of time. Chemicals, employed properly to provide a balanced diet or as supplements for the plant without leading to soil damage, can increase the amount and quality of food production in many areas, however.

Proper soil management is the key to successful gardening. Once you learn how to build fertile, loose soil, you can expect to have the best garden in your neighborhood. Without good soil you will have only a mediocre garden, even if you add sufficient nutrients. Unless you develop an intensive method of soilless culture, you will need as much organic matter in your soil as you can get. Even the most avid organic gardener rarely obtains as much as he wants. Since bacteria decompose it, it constantly disappears and needs replacement. If you continue to cultivate and fertilize the soil without adding any organic matter, it will get more difficult to handle each year. Clay soil will harden to brick, and sandy soil will turn to pure sand that cannot retain either moisture or nutrients.

How can you get enough organic matter into your

soil? You will need to exploit every possible source. If your land is small, you will have to import most of it either as manure and hay that you purchase, or leaves, grass, and refuse that you collect from your neighbors. But if you do have enough land, let half of it produce food for you while the other half grows clover and grass for cover crop. Then you can gradually build up the organic matter by crop rotation—the method commonly used by large-scale farms who must compost in the field instead of in a heap. Poor soil might even need fertilizing to get a cover crop started, but once you start plowing under a good stand of clover, the soil will begin to build itself. Remember that legumes like clover and vetch will add extra amounts of nitrogen by means of the nitrogen-fixing bacteria in their root nodules. All legumes are good soil builders, so use them as much as possible.

Real satisfaction comes from watching the soil improve year by year instead of gradually wearing out. Some farmers mine the soil instead of managing it when they harvest a crop each year but return nothing to improve it. Once you have a piece of ground built up to a rich, loose loam, you can keep it that way. It will hold nutrients and moisture like a sponge, conserving both for your crop. Rich soil can produce a bigger and better crop with less loss of water and nutrients. Once the soil deteriorates, it requires large amounts of water and nutrients to raise even a poor crop.

So you see, adding organic matter to your soil is like putting money in the savings account. It not only improves with each addition, but also has a greater capacity to produce for you.

CHAPTER 3

PACKAGES OF LIFE

A seed is a marvelous creation. A hard, dry, lifeless-appearing bit of material, it contains a living embryo with the potential of becoming a fruitful plant. However, it will never do anything unless placed in the right circumstances for growth. Here is where the gardener comes in. Growing plants from seed is a rewarding experience if it works, but most frustrating if it fails.

A seed is really a living plant all wrapped up in a special package that protects it until time to germinate. The cells of the embryo plant carry on all of the normal life processes. The seed stores enough food to keep the embryo nourished during its long dormant period and to supply its needs until it develops into a green plant able to produce its own food. When you enjoy a slice of bread, remember that the flour comes from the nourishment accumulated in the grain for the embryo until it forms green leaves to manufacture food for itself.

If the embryo runs out of food before it becomes a green plant, it will die. You may have read stories about barley seeds, taken from the wrappings of an Egyptian mummy supposedly three thousand years old, that grew when planted. Researchers have learned that archaeologists had packed the mummies in barley straw for movement from the tombs to the museums, and the barley seeds were actually only a few years old when planted. Scientists have performed some experiments to see how long seeds do

live. One group buried containers with a variety of
seeds in them as early as 1879. Over a period of seventy years, they dug up the seeds one bottle at a time
and tested for germination. Some seeds lived for only
a few years, but at the end of the seventy years three
kinds still grew when planted.

If you wish to save seeds from some of your favorite plants and keep them for several years, you must
store them properly to preserve life in them. A seed
kept cool will not use up its food supply as rapidly
because the lower temperature greatly slows the life
processes. Ideally you should store seeds at 40° to 50°
F, and if that is not possible, keep them at least below
60° F. As soon as a seed warms up, its biological
activity begins to speed up in anticipation of germination. For a seed to grow, it must have water. A seed
has little water in its tissues, but as soon as moisture
enters, it begins to soften and swell. To make seeds
remain dormant, you must keep them dry as well as
cool. In damp climates you may need to put them in
tight containers. Most seeds will remain viable for
many years. Only a few are so fragile that they die
during the first year of storage.

What if you have some seeds that you have kept for
several years? Can you tell if they are still alive before
you go to all the work of planting them? It may be a
crop that you want to start at a specific time. If you
take a chance on an old seed, and it doesn't grow, then
you are way behind schedule. In testing seeds, gardeners have used for many years the rag-doll method,
so-called because you roll the seeds up in a damp rag
and place them in a warm spot. Unroll the damp rag
each day to see what the seeds are doing. Most will
begin to swell within a few hours, and after a day or
two you should notice some evidence of sprouting.
Some seeds may require as long as a week for germination, so you must keep the rag wet until the germination period ends. A plastic bag may help prevent

the rag from drying out rapidly. Or you may substitute a paper towel for the rag—in fact, any arrangement that your ingenuity contrives will work as long as you remember that the seed needs moisture and warmth. By counting the seeds that sprouted and comparing that with the number of unsprouted seeds, you calculate a percentage of germination. If only half of the seeds germinate, you will need to plant them twice as thick in the row to get the number of plants you desire.

Large seeds like beans and peas and corn must soak up considerable amounts of water before they can sprout. It may take several days for the seed to absorb enough moisture to start the germination process. You can speed things up by soaking large seeds overnight before planting them. Just place them in a dish of water several times the volume of the seeds. Don't try it with fine seeds like carrots and lettuce, however, or you may never be able to spread them out in the row, since they will all stick together. Fine seeds require planting in relatively dry, loose, crumbly soil and then watering thoroughly.

Remember that in addition to moisture, seeds require warmth for germination. Some demand much warmer temperatures than others. Melon and tomato seeds will not germinate in cold soil even though they have sufficient water. Eager spring gardeners often plant melon, squash, and cucumber seeds in the cold dirt, hoping that the warm sun will perform the necessary magic to bring them up. Not until the soil itself is actually warm will such seeds germinate, and the cold spring rains will likely keep them from growing for so long that they rot. Nature has a good reason for delaying the germination of some seeds until warm weather arrives. Melon and squash seedlings are susceptible to frost damage and seedling decay or damping-off. Seeds slow to germinate may escape the late frost. When they do sprout, the weather has turned warm enough to keep them growing fast so they can get ahead of decay organisms in the soil.

Gardeners are never content to let nature take its normal course, but they always try to improve conditions in hopes of a bigger and better crop. Now that you understand the requirements of a germinating seed, you can devise your own little tricks. Should spring seem too slow to suit your schedule, you can hurry the soil-warming process with a sheet of black plastic over the planting area to soak up the sun's heat. The seeds will then have an extra boost to get them up quickly. As soon as the seedlings appear, however, remove the black plastic to allow the sun to shine on the green leaves.

And should winter still persist, you may have to cover the tender plants with boxes or hot caps on cold dull days and frosty nights. Of course you can also start fragile seeds in a greenhouse, a hotbed, or even on the windowsill. By timing it just right you will have vigorously growing tomato, pepper, eggplant, or even cucumber, squash, and melon seedlings ready to transplant into the warm summer soil as soon as the danger of frost passes.

Other seeds call for more than just moisture and warmth to germinate. Have you wondered about the strange instructions on the package of certain kinds of sweet peas? Sometimes they suggest placing each seed in the water-filled compartments of the ice-cube tray and placing it in the freezer. The seeds remain frozen into the ice cubes for a few weeks before you plant the ice cubes, each with a frozen seed inside. Such seeds contain chemical growth inhibitors built into them that freezing temperatures will alter. Not until the growth inhibitors break down can the seed germinate. Some wild plants would quickly become extinct if the seeds they produced in late summer all germinated with the fall rains. The severe winter would destroy all the delicate seedlings. The chemical inhibitors will not allow the seeds to sprout until *after* the winter freezing period.

The germination pattern of a wild shrub, the western redbud, intrigued me a few years ago. Seeing the abundant crop of seedpods on the redbuds along the highway, I decided to collect some and grow my own redbud plants for my yard. Carefully I placed the seeds in good soil and kept them watered. After several months went by with nothing happening, I dug them up and found them the same as when I planted them. They hadn't even begun to swell. A little examination under the magnifying glass showed me why. The shiniest, hardest wax I have ever seen covered each one. If a housewife could have a coating of that wax on her floor, she probably would never have to polish it again. The seeds were completely impervious to water, never sprouting until something removed that wax.

In the natural state, the shrubs grow in areas where brush fires are common. The fire will burn up the redbud shrubs and the leaves on the ground, but many of the seeds will only get hot enough to melt the wax from them. Then when the rains come, they soak up moisture, sprout, and grow a new redbud shrub where fire wiped out the old one. Should all the redbud seeds germinate every spring, a fire might kill the living plants at a time when the plants had no seed and could not replace themselves.

Some seeds will not sprout unless exposed to light. This might help you understand why weeds continue to grow in your garden each summer, even though you have kept them cleaned out faithfully for several years. Every time you cultivate, you bring some more weed seeds to the surface, and the light triggers them into activity.

Watching a seed grow will help you understand better what makes for good germination. Radish seeds respond easily and quickly. Place them on a

damp paper towel on a plate and cover them with a
large glass or bowl. If you give enough moisture and
keep them at room temperature, they will begin to
swell within a few hours and sprout overnight. The
first thing you see is the root. A magnifying glass will
reveal the delicate root hairs. The root must develop
first to support the tiny plant as it grows from the
embryo. The first leaves are the fleshy halves of the
seed. They contain food for the plant until it can reach
the surface of the soil. As soon as that happens, the
plant develops green leaves and manufactures food by
photosynthesis.

A seedling does best in soft loose soil with plenty
of moisture and air. Till the soil six to eight inches
deep and break up the large clods. Cover the seed
with a thin layer of soil. Water it gently with a fine
mist so you don't wash the seeds out or make the soil
muddy. Too much water will drive out all the air and
compact the dirt tightly around the seed. If it doesn't
suffocate, it still may not manage to break through the
hard crust that forms on the surface.

How to improve the garden soil to make a proper
seedbed depends a lot on what kind you have to start
with. If you are fortunate enough to have soft sandy
loam with lots of organic matter in it, and you can
loosen it while it is still damp, your planting job is
easy. Make a shallow depression for your fine seeds
and cover them one-fourth inch deep to prevent them
from drying out. If the soil particles touch each other,
they will bring up the moisture from below like a
wick. You shouldn't have to water until the plants are
up, and by that time the rain may have done it for you
anyway.

If you live in an area of heavy rainfall, leave the soil
loose so the rain can trickle down through without
puddling. With heavy clay soil which forms a hard
crust on the surface after watering, you may wish to
put a thin layer of screened compost or a mixture of 50

percent peat and 50 percent sand over the seeds. It provides a soft medium for the seedlings to push through and protects them from drying out. The organic matter you add to your soil you should mix months ahead of planting time so the bacteria will have had time to break it down into its chemical components. Bury concentrated nutrients like chicken manure or chemical fertilizer several inches under the row so they will not burn the tender root of the seedling.

Large seeds like corn and beans you can plant deeper than fine seeds because they have enough food stored up to bring the seedling to the surface before it runs out. However, under good soil conditions, even corn and bean seeds can tolerate only one inch of soil. In more than that, the warm sun can't reach them to hasten sprouting.

Remember that the seedling has a small root system at first, so keep the soil moist. Don't let the surface remain wet too long after the seedlings appear, though, or they may send their roots out only in the top layer of soil. Then the first real warm day that you aren't there to water the garden, disaster will strike. The top layer of soil dries out and the plants all wilt. They don't have a root system into deeper soil where the moisture is. Force the seedlings to send their roots down by withholding water gradually at first. Even-

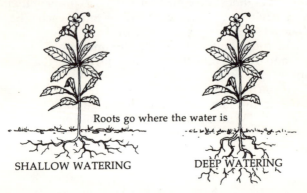

Roots go where the water is

SHALLOW WATERING   DEEP WATERING

tually you should be able to let the top inch of soil dry out before watering again. The roots will grow toward water wherever it is. A deep, extensive root system makes a much healthier plant because it will have a more constant supply and better access to nutrients.

As we have pointed out before, the only parts of the root system that can take up water and nutrients are the root hairs. Root hairs are merely extensions of thin-walled cells on the sides of the delicate new root tip. They are so small that as many as one hundred may occupy an area the size of a pinhead. However, an extensive root system like that of a four-month-old rye grass plant may have as many as 15 billion root hairs with a total surface area of 3,000 square feet. They provide for a rapid uptake of water on a hot dry day to keep the plant from wilting. The water passes through the conducting system to the leaves, which use it in photosynthesis. Large quantities evaporate through the microscopic pores in the leaf surfaces—a process called transpiration. It helps keep the water moving and also cools the plant. A large corn plant transpires four gallons of water on a warm day. One acre of corn uses 8,400 gallons of water in a day.

As the root tip uses up the moisture in one area, it grows into a new one. Some plants will send roots several feet deep to find more water. The root system must equal the leaf system. If a plant grows many leaves and then runs out of water, it will not be able to supply the cells. The leaves will then droop or wilt. You can save a plant that has just started to wilt by giving it extra water. Some plants like squash that have many large succulent leaves will manage nicely during the cool part of the day. But when the hot sun hits them, they don't have a root system extensive enough to pump water as fast as it transpires. The leaves droop even though you have just watered the plants. But when evening comes, the roots can catch up again.

Understanding the water system of a plant will help you know how to care properly for transplants. When you take a small tomato or cabbage plant from its container and set it out in the garden, you are bound to disturb the root hairs. No matter how much water you pour on it, the plant will not have enough root hairs to cope with its water needs. Pinching off half the leaves will cut the water loss by transpiration. Also you could shade it from the direct sun, or better yet, choose a cool, cloudy day for transplanting. Some of the starting solutions available at the garden store contain substances that encourage root growth.

The tiny root system of vegetable transplants purchased at stores is a real handicap to proper growth. Grown from seeds in a warm, moist greenhouse where a small root system can support heavy foliage, they receive a severe shock when placed directly into the garden. I have found that such plants will make much greater progress if first transplanted into four-inch pots or milk cartons and kept indoors or in a cold frame until they develop a bigger root system. It also delays planting time for such tender plants as tomatoes until the soil has warmed up and the danger of frost has passed. Plants handled in this way jump way ahead of those set out directly into the garden.

CHAPTER 4

WHICH VEGETABLES
SHALL I GROW?

One of the most enjoyable winter activities is sitting by the fireside with a stack of seed catalogs, studying the possibilities for the coming season. It is an important part of gardening, since only by trying new things can the experienced gardener satisfy himself that the old tried and true is still the best. How do you decide what to grow? If you rely on the descriptions in the seed catalog, you had better plan to start a seed store of your own, because they portray every variety in glowing terms as the largest, tenderest, tastiest, and most productive.

Actually you should consider at least five factors when selecting seeds for your garden. First, what climate do you live in? The length of the growing season will dictate to a large degree the kinds of vegetables that will thrive. However, the gardeners of southern California and Florida have problems as well as those of Washington State and Maine. The seed producers and plant breeders have bred varieties particularly adapted to different climates. The catalogs will sometimes specify which ones are best for yours. Be sure to compare notes with other gardeners in your area. Experience is the best guide. Don't be afraid to try something new each year, but make it a small part of your crop so that if it fails, you won't lose out completely.

A second important factor in determining what to grow is that of room. How large is your garden space? Can you expand, or are you limited to a small

47

	Distance Apart		Amount of seed, or no. of plants for 50-foot row
Kinds	Rows, feet	Plants in rows, inches	
Bean, bush	2	2 to 3	4 oz
pole	2	8 to 12	4 oz
Beet	1½ to 2	1 to 3	½ oz
Broccoli, early	2½	18	1 pkt
late	2½	18	1 pkt
Cabbage, early	2½	18	1 pkt
late	2½	18	1 pkt
Carrot	1½ to 2	1 to 2	¼ oz
Cauliflower	2½	18	1 pkt
Chard, Swiss	2	8 to 12	½ oz
Chinese cabbage	2	12	1 pkt
Corn	2½ to 3	12 to 18	2 oz
Cucumber, slicing	4	12 to 24	⅛ oz
pickling	4	12 to 24	⅛ oz
Endive	2	8 to 12	1 pkt
Herbs	2	6	1 pkt
Lettuce, leaf	1½	6	1 pkt
head	1½	12	1 pkt
Muskmelon	4	12 to 24	1 pkt
Onion transplants	1½	3	1 pkt
seeds or sets	1½	2 to 3	1 pkt seed, 1½ lb sets
New Zealand spinach	2 to 3	24 to 36	1 oz
Parsnip	1½ to 2	2 to 4	¼ oz
Pea	1½ to 3	2	4 oz
Pumpkin	6 to 8	36 to 48	1 oz
Radish, spring and fall crop	1	1	½ oz
Rutabaga	2	6	½ oz
Spinach, spring and fall crop	1½	4 to 6	½ oz
Squash	6 to 8	36 to 48	1 oz
Tomato, staked	2	18 to 24	25 to 33 plants
unstaked	3	36	17 plants
Turnip	1½ to 2	3 to 4	½ oz
Watermelon	4 to 6	12 to 24	½ oz

Depth to cover, inches	No. of days seeding to harvest	Approximate yield per 50-foot row	How to use or store
1½ to 2	52 to 70	30 to 50	Fresh, fresh frozen, canned,
1½ to 2	65 to 75	qt	pickled
½	55 to 70	250 roots	Fresh, pickled, canned, cool cellar
Transplants	60 to 80	30 to 40	Fresh, fresh frozen
½		qt	
Transplants	60 to 80	30 heads	Fresh, raw
½	100 to 105	30 heads	Fresh, raw, kraut, or storage
½	60 to 75	30 to 75 lb	Fresh, raw, canned, cool cellar
Transplants	60 to 80	30 heads	Fresh, fresh frozen
½	50 to 60	Use all season	Fresh
½	70 to 90	50 heads	Fresh
1 to 2	70 to 100	45 to 75 ears	Fresh, fresh frozen, canned
½ to 1	65 to 75	100 to 150	Raw
½ to 1	60 to 70	50 to 150 fruits	Pickled
¼	70 to 90		Salad
¼			
¼	40 to 50	100	Raw
¼	70 to 75	50 heads	Raw
½	70 to 100	75 to 150	Fresh
Transplants seed ½, sets 1	115 to 135 95	50 to 75 lb	Raw, fresh; dry, dark, cool storage
½	60 to 80	Use all season	Fresh
½	120 to 150	150 to 300 roots	Store sand, moss, sawdust; or leave in ground over winter
1½ to 2	60 to 80	20 to 40 qt pods	Fresh, fresh frozen, canned
1	110 to 130	30 to 50 fruits	Fresh, stored dry
¼	25 to 35	30 to 100 bunches	Fresh
¼	110 to 130	100 lb	Fresh, stored
½	40 to 45	1 to 2 bu	Fresh, fresh frozen
1	90 to 115	100 fruits	Fresh, stored dry
Transplants	100 to 130	150 to 300 fruits	Fresh, canned
Transplants	100 to 130	150 to 300 fruits	Fresh, canned
¼	50 to 70	150 roots	Fresh
1	90 to 100	75 to 100 fruits	Fresh

backyard plot? Experts have demonstrated that 300
square feet (15 feet by 20 feet for example) provide
enough space to grow all of the fresh vegetables that a
family of four can consume. However, you may desire
more than just fresh vegetables in season. You may
want corn, beans, and potatoes enough to store away
for the winter. If you do not have much room,
remember that root and leafy vegetables produce
more food per square foot than others. Such bountiful
foods include carrots, turnips, radishes, beets, lettuce, chard, cabbage, and spinach. Vines such as
squash, cucumbers, and pumpkins require the most
space. Sometimes you can train them on a fence.
Tomatoes, cucumbers, beans, and even squash will
grow vertically instead of horizontally.

The third factor in selecting vegetables involves
how much you can use. Inexperienced gardeners frequently plant equal amounts of each kind. Compare a
fifteen-foot row of carrots with a similar-length row of
radishes. The carrots grow slowly and will not reach
harvest until mid or late summer. You may pull them
from the garden throughout the fall and even into the
winter in milder climates. The radishes, on the other
hand, develop quickly and require harvesting within
a week or two after the first one is ready to eat. They
quickly go to seed, and the roots toughen and become
hot. How many radishes can you eat in two weeks?
Probably a three-to-four-foot row would suffice. Also
you can make small plantings a week apart for several
weeks and spread the harvest out somewhat.

Home gardeners also frequently overplant lettuce.
A few feet of vigorously growing leaf lettuce will make
many crisp salads. Overplanting results in a waste of
time, space, and energy, and it lends to disappoint-
ment when the plants get bitter and go to seed. Swiss
chard will continue to send up tender new leaves for
four to six weeks. A small row will provide an abundant harvest if properly cared for.

A fourth factor may be your likes and dislikes. While you may enjoy watching a big patch of okra and eggplant thrive, unless you like to eat a lot of them, you might better have planted something that suits your taste buds. If you put in a whole row of zucchini squash, you had better plan to eat nothing but zucchini squash when it starts to produce because you must pick it almost every day. It's a lot of fun to see pumpkins on the vines in the fall, but how much do you eat other than for pies? A few people like boiled pumpkin, but most prefer some variety of squash. But if you learn to make "pumpkin" pies from squash, then it can serve both purposes. I have even seen most interesting jack-o'-lanterns made from squash.

The fifth factor may be one of the most important. What kinds of vegetables produce the most food with the least amount of time and effort? Swiss chard would probably stand near the top of such a list, and perhaps asparagus and lettuce also. Experience will reveal which kinds of vegetables offer the most success in your particular setting. Some varieties of beans will bear over a long period of time. Once they start, from then on it is mostly just a matter of watering and harvesting to keep a good food supply coming to your table.

One old-fashioned variety of bean that we always called cranberry beans you can pick first as green snap beans. Then as they mature, you can shell them out as rich, meaty butter beans, and whatever you don't use that way will ripen and dry in the pods. Store them for winter as dry beans similar to pinto beans. I have discovered that the more modern but unappealing name for cranberry beans is now horticultural bean. Don't let the name turn you away. You can get more yield from a three-way bean like that for the time and energy expended than from three separate varieties. Some varieties of lima beans will continue to produce over a period of several weeks, giving you fresh green limas on your table for many meals. Some years I pick green limas until the frost kills the vines. If too many develop for our needs during that time, we either freeze them green or allow them to ripen and dry for winter storage.

Leafy crops such as lettuce, chard, spinach, cabbage, and broccoli are cool-weather plants which you can start early in the spring. They grow quickly and soon reach the eating stage, because you do not have to wait for fruiting season. Cool weather requires less watering, and insects don't multiply as rapidly then. However, since it is leafy growth you want in such plants, you must make sure the soil is rich in nitrogen. Also, provide abundant water because once they stop growing, they are hard to start again. Grow them in spring and fall, because the long summer days will influence them to go to seed.

The warm-weather vegetables are those that produce fruit, like tomatoes, melons, eggplant, and beans. Such plants will not develop during cool weather, as they are susceptible to frost damage. Besides requiring a longer growing season to mature and ripen their fruits, they also need a better balance of nutrients in the soil, since they must produce more than just leafy green growth.

Here are some ideas to help you plan your garden to meet your own needs:

Vegetables that are easy to grow:

Bush beans	Leaf lettuce
Beets	Radishes
Carrots	Squash
Cucumbers	Swiss chard
Kale	

Vegetables that yield most for space and effort:

Bush beans	Lettuce
Beets	Tomatoes
Carrots	Zucchini

Vegetables that grow best in cool weather:

Broccoli	Spinach
Kale	Turnips
Lettuce	Peas
Radishes	

Time from planting to harvest:

Radishes—22 days	Turnips—45 days
Mustard greens—35 days	Green beans—45-50 days
Lettuce—38 days	Peas—55 days
Spinach—42 days	Swiss chard—60 days

Start seedlings indoors six to ten weeks early:

Broccoli	Eggplant
Cabbage	Peppers
Cauliflower	Tomatoes
Celery	

Most vegetables thrive best in full sun with a good supply of water and nutrients. Make sure corn doesn't shade the other plants in your garden. Since it also needs a lot of room, don't crowd it. Most garden books recommend that you sow corn in hills, three plants to a hill with the hills 18 inches apart in the rows and the rows 30 inches apart. Don't forget that corn is wind-pollinated, so employ a block of several short rows rather than a long single one.

Your vegetable garden will grow best in soil that is near neutral, not acid or alkaline. Soil with a lot of compost and organic matter will usually regulate itself because organic matter works like a sponge to take up excess acidity or alkalinity. You can do a more accurate job of adding fertilizer to your garden if you have the soil tested. Whether you employ organic or nonorganic fertilizers, avoid overfeeding. Heavy soils will retain plant food longer than light sandy ones. A number of people have learned to grow beautiful vegetables in containers with no soil at all, only sawdust and sand to which they add a carefully measured balance of chemical plant foods. While the method takes some skill in measuring and following directions, if done correctly the results can be remarkable. All plants get the same amount of food and water, and they will grow rapidly and evenly.

It is a good plan to rotate your planting pattern each year so you don't put identical vegetables in the same spot year after year. Each plant has individual needs, and you will not deplete the soil as rapidly if a different plant grows in it each year. Some plants such as the legumes will add nitrogen to the soil. If you rotate them each year, the entire garden will eventually benefit from their growth. Some plants harbor certain pests and diseases. Raising them in the same soil several years in a row may lead to their destruction by their natural enemies.

What do you do if suddenly one morning you discover some insect pest has invaded your garden and threatens your vegetable crop? I suppose you will find as many different reactions to such a problem as there are people with gardens. Let me make one simple suggestion to start with that should help everyone, whatever method he chooses. If you can discover what kind of insect is doing the damage, you will most likely have greater success in choosing your weapons to fight it.

Not everyone can take a class in economic entomology, but the many books and pamphlets available give drawings and descriptions of the common insect pests. If you can't find out that way, don't forget that the county agricultural agent or the agricultural extension service in your area has already been paid to help you, so take advantage of their services. You don't have to follow their prescription if you don't want to, but you can certainly gain from their knowledge and experience. As a result, you might save a lot of time and money trying to eradicate something that isn't doing any real damage to your garden.

What about biological control as opposed to chemical control of insects? It seems that whenever they have a choice, most gardeners would prefer to avoid chemical pesticides. Hopefully you are gardening in a complex ecosystem where their own natural enemies will hold insect pests in check. You cannot develop the ideal biological control system, though, on a moment's notice, or even in one growing season. Those not fortunate enough to live in one may need to begin a long-range program to establish their own little version. But what will you do in the meantime? Will you have to stand by and watch while ravaging insect pests devour your choice vegetable plants? Fortunately, you can do some things to protect your plants while you wait for nature's balance to take over.

No matter what you do, protect the beneficial insects in your ecosystem. To recognize them from the harmful ones, you may need to do a little studying; also ask for help. The intricate patterns of activity I see on my little acre have fascinated me over the years. Opening your eyes to the panorama of life going on in your own backyard will both entertain and educate you at the same time. As you learn more about the various forms of life in your garden you will be better prepared to guide the system in the direction of more food production for yourself. More helpful insects exist than harmful ones, so you will probably learn to recognize the latter first. If you have a small garden and lots of time to spend in it, you might regulate pests simply by checking your plants daily and removing the pests by hand or with a stream of water.

On the other hand, you may have too large a garden for simple hand picking or washing. Yet you may still achieve some control without destroying the beneficial insects. An insecticide that destroys plant pests at the time you spray and then disappears within a short time will not damage the beneficial insects as much. Perhaps you may wish to use one of the botanical insecticides such as rotenone, pyre-thrum, or nicotine, which are effective poisons, but they break down and do not last long. Some of the organic phosphate chemicals such as Malathion are also biodegradable. Whichever insecticide you decide on, read the label carefully and follow directions. Too many people follow an old adage, "If a little is good, a lot is better." Dosage beyond that necessary will do nothing but harm.

By avoiding the persistent chemical insecticides like DDT, chlordane, and other chlorinated hydrocar-bons, you may be able to hold harmful insects in check but not wipe out the beneficial ones. Increasing the complexity of your ecosystem by planting a vari-ety of plants around and through your vegetable gar-den will establish more stability. By learning enough about your ecosystem, you may eventually reach the ultimate goal of natural control. Until then you may need to keep two things in mind. First, a program that utilizes as much natural or biological control as possi-ble with the assistance of biodegradable insecticides properly applied can produce a safe crop of food with a minimum of harm to the ecosystem. Second, you can tolerate a certain amount of insect damage as long as your garden produces the food you need.

The idea that the gardener must completely eradi-cate insect pests to be successful will prevent a stable ecosystem. For instance, if you want ladybugs to hold the aphid population in check, you must have some aphids for them to eat, or they will move to a place where aphids do exist. While it doesn't mean that you need to place aphids on your plants to attract the beetles, still you must not completely eradicate them with an insecticide unless you plan to continue doing so all season without any help from the beetles. On the other hand, if the aphids get a head start before the ladybugs arrive, use a short-lived insecticide until the beetles can take over.

CHAPTER 5

SMALL FRUIT PRODUCTION

Some kinds of fruit can grow in almost every climate, but the usual question arises as to what varieties are best Tree fruits are a long-term investment that takes a lot of space. Small fruits are more feasible for the gardener, and they will yield a crop more quickly. Berries are the most commonly grown small fruits.

Strawberries are perhaps one of the easiest to start with. Plants set out in the late fall or early spring will bear some fruit the first season. Several varieties will produce ripe berries until the first frost. They do not always grow the biggest and best berries but supply fruit during a much longer time. Other varieties are one-crop strawberries, which produce a large harvest in early summer and then rest for the remainder of the year. Commercial growers do best with one-crop berries. Most home gardeners like to have more than one type in order to take advantage of the good points of several different ones. The catalogs list many varieties, each with its own particular features. It is difficult to decide which suits your needs best, so you may have to try a number of them before you settle on the one or two that you prefer.

The strawberry, relatively hardy, will grow almost anywhere but performs best in rich soil with abundant moisture and cool weather. The soil must be well drained, though, because the roots will rot in soggy soil. You can usually purchase bare-root runner plants in the fall or spring. They are the small plants at the end of the stringlike stems that spread out from a plant during the summer. Usually gardeners set them out about one foot apart in the row and two to three feet between the rows. You will have to keep the runners cut off, or they will fill in the space between rows, leaving no place to walk while picking the berries. Some people let the runners grow within the row or even for a foot or so on either side. It makes a solid bed of plants that they can reach from either side by the narrow path in between. You need room to walk not only for picking the berries but for pulling weeds and other maintenance activities.

Strawberries do best in soil rich in organic matter. Work large amounts of straw and barnyard manure into the soil a few months before planting. You may turn in a cover crop just before planting, but be sure it doesn't produce so many weeds that they crowd out the strawberries. Young runner plants will have roots from three to six inches long. Do not stuff them down into the small hole and pack them in, because strawberries tend to be shallow rooted. Spread the roots out in a shallow trench and cover them with two to three inches of loose soil. Don't bury the crown of the plant or leave it too high up in the air. Look at the runner plants before you dig them up, and you can see how the crown is at the soil surface.

You can use runner plants from previous plantings as long as they remain free of diseases. With so many new varieties coming out all of the time, you will probably buy some new plants almost every year anyway. Not planting them on the site of the old patch, if you can avoid it, will help prevent the spread of disease. Most types of strawberries will continue to bear a good crop for three to four years from the same plants. If you let new runner plants develop between the old ones, you can remove the old plants after two to three years.

The continually renewed patch will bear as long as

no disease arises. Should you suspect some infection, ask your county extension office to advise you on how to control it. Sometimes the best way to avoid disease is to buy new plants and set them out in ground where strawberries have not grown before. Rotating a strawberry patch from one part of the garden to another can improve your soil, particularly if you add large quantities of organic matter to the patch. A continuous mulch of two to three inches over the roots of the strawberry plants will keep them cool and moist, especially important during warm dry weather.

Sprinkling strawberries often damages the ripe berries. You may be able to use your ingenuity to devise a method of watering that does not get the berries wet. Some of the new drip irrigation systems might give you a start.

Blueberries are more restricted to cool climates than are strawberries. They are also more particular about soil conditions. If you have wild blueberries in your area, observe their growing habits and then try to match them as closely as you can for the domestic varieties in your garden. Acid soil is one of the most important requirements for blueberries. Usually you can achieve it by adding large amounts of peat moss or sawdust before planting. Sulfur will increase the acidity. Ammonium sulfate will accomplish the same thing and add some nitrogen at the same time.

Since blueberries are adapted to a cool climate and are shallow rooted because they usually live in boggy areas, you must keep the roots well covered with a thick layer of mulch. It will protect the roots from drying out and also help to keep them cool. Don't make the mistake of using pure peat moss as a mulch. Should it ever get really dry on the surface, it will shed water like a roof. As a result, you may not be able to get water to the root system before damage results. Mix about 40 to 50 percent sand or soil with the peat moss and then use shredded leaves, bark, or coarse compost over the surface to keep the peat-sand mixture from splashing during sprinkling. The sand in the peat moss provides a passageway for the moisture, and it will take water more evenly.

By selecting several different varieties of blueberries, you may get them to ripen over a period of six weeks. Some types have an acid and tart taste while others are much more mild. Several turn soft when ripe and may bruise from handling, while others remain firm. In addition to providing delicious fruit, a few blueberry bushes will add some striking fall colors to your yard or garden. Here again they differ in shades and intensities of red to purple colors of the leaves. It is surprising how widely blueberry growing has spread over the country. Unless you live in a hot, dry southern climate or an area of alkaline soils, you should give blueberries a try. You might be pleasantly surprised.

Cranberries are even more restricted in their distribution. Unless you have a permanent peat bog or a low-lying area that you can flood, don't waste your time. The Pacific Northwest and the New England states are the major cranberry-raising areas. I remember as a child picking wild cranberries where you had to keep moving or you would find water over your boots. The trailing stems of the wild cranberry plants grew over the surface of a peat bog that was only about two feet thick with several feet of water underneath. The cranberries were small and probably not as tasty as the domestic varieties, but they had an extra special flavor to a young lad who had spent several hours gathering a small pailful.

Wild huckleberries may not have the same flavor as domestic blueberries, but where available for the picking, they make an excellent addition to the year's supply of fruit. In the Pacific Northwest, they cover thousands of acres of public lands such as national forests. Vacationers and weekend hikers harvest only a small fraction of the yearly crop.

Domestic brambleberries such as blackberries and boysenberries offer another nice source of fruit. Most varieties of blackberries are relatively easy to raise and thrive throughout most of the United States. Once a blackberry establishes itself, you will need to restrict its growth rather than promote it. Left unattended, blackberries soon form an impenetrable mass of thorny brambles. Only the outer layer of a thicket produces berries. You can obtain an equal quantity of fruit on a trained vine that you can cut back each year. Pruned faithfully each year, blackberries are not hard to manage, but once a tangled thicket develops, most people give up in discouragement.

The berries appear on the cane growth from the previous season. As soon as you pick the berries, cut back to the ground the cane that bore them. Most varieties will have already started the new growth for the next crop, so remove only the canes that have overwintered and leave the new ones. When winter arrives you should have a well-developed system of canes tied to wires or draped on a trellis. They remain dormant until spring, then produce blossoms and fruit.

If you wish to avoid vicious thorns, try the thornless boysenberries. Some say they don't taste as good as the thorny ones, but you may want to choose between scratched arms and lesser flavor. Blackberries respond to fertilizing. Even though it means more work to handle longer canes, your yields will proportionately increase. A layer of well-rotted manure applied during the winter will provide necessary nutrients and also form a protective mulch for the root system.

The wild blackberries in some areas, having had no particular care, are frequently difficult to pick and are of inferior quality. In many places the "red-berry disease" has almost eliminated them as food. It is an infestation of a microscopic mite that lives in the buds over winter and develops within the ripening berries. The berries do not ripen properly. A nice-looking berry may have only a few "kernels" or druplets that remain red, but that is all it takes to make it so sour that it will twist your mouth. A winter spraying of insecticide will control the mite, but you must thoroughly and consistently apply the treatment for it to be effective. Most people prefer domestic varieties and leave the wild ones for the birds wherever the red-berry mite exists. Thirty-five years ago we had never heard of the red-berry mite, but regions that once yielded tons of sweet ripe berries now are almost worthless because of it. If you live in an area where the wild blackberries are still good, be thankful and take advantage of them as long as you can.

It seems the wild blackberries that are the hardest to get have the nicest flavor. I remember as a child many hours spent scrambling over brush piles, fallen logs, and stumps in search of the small trailing wild blackberries. A long afternoon in the hot sun was well worth the delicious sauce, jams, and pies we made from those little berries that seemed to have twice as much flavor as the large evergreen blackberries. Sometimes we would find a small patch of black caps, the wild black raspberries, and they would hasten the slow process of filling our pails. The black caps added a special taste that increased our pleasure when eating the hard-earned prize. Let's hope the day never comes when pests and diseases so riddle wild berries that we can no longer enjoy them.

Raspberries can be a satisfying fruit crop for even the small garden. As they grow mostly vertical, you can plant them next to a fence where they will not take up much space. Like strawberries, red raspberries consist of both one-crop varieties and everbearers. The one-crop varieties ripen in the late spring or early summer. The everbearing varieties have both a spring crop and a fall crop. In warm climates the fall crop may be more abundant and of better quality. We have picked large juicy berries from a September Red variety right up to the first hard freeze. The cooler autumn weather seems to favor larger berries that ripen more slowly and don't tend to dry out so quickly.

When raspberry canes die after the spring crop, trim them back to the ground as soon as they finish bearing. Watch out for the new canes, though, because they will produce the fall crop, if it is an ever-bearing type, or the spring crop of other varieties. The canes with autumn berries will drop their leaves after freezing weather comes and stand like so many dead sticks through the winter. Far from being dead, though, they will send out fresh green leaves and clusters of blossoms early in the spring. The top 6 to 12 inches of those canes that previously bore fruit will not put out new leaves, and you can prune the dead tips off as new growth starts on the rest of the cane. While not essential, it does improve the appearance of the patch.

Unlike strawberries, raspberries do not send out surface runners, but the roots grow outward from the base just under the soil surface. New plants sprout up from the lateral roots. If you don't keep cutting them down, you will soon have a 6-foot-wide jungle instead of a neat row that you can pick from both sides. Those sprouts make an excellent source of new plants, however, when you are ready to start additional rows. Since raspberries renew themselves every year, they don't need replanting as frequently as strawberries. You will find the raspberry row less demanding to care for and the berries easier to pick if you have a wire on which to train them. Either tie them to one central wire stretched tightly between posts or support them loosely between two wires a foot apart.

Some varieties of raspberries are particularly susceptible to a virus disease. But the virus-resistant strains available now will produce consistently even in infected areas. A few insect pests attack raspberries, too, but if you maintain the plants in a healthy condition, they can usually outgrow the damage. Raspberries respond well to fertilizing, and a rich soil kept up by yearly applications of fertilizer will result in the most healthy plants and the nicest crops.

Grapes are another fruit you should consider. Warm climates ensure the most abundant and varied crops, but by selecting those varieties developed especially for the north, you can grow grapes throughout most of the United States. European grapes, including Thompson seedless, do best in warm southern climates. Not only do they freeze out in the northern winters, but they rarely get fully ripe in the shorter-growing seasons. Nothing more disappoints a grape enthusiast than to have a vine hanging heavy with beautiful clusters of grapes that never fully sweeten. The American grapes such as Concord and Niagara are more satisfactory in northern regions. Hardy, they ripen earlier than most European types. If you study the catalogs you will find some interesting varieties designed for most climates.

Deep-rooted plants, grapes do not require surface watering. In dry climates they send roots down to permanent water sources and can grow 20 to 30 feet of canes right through the heat of summer. Such a trait makes them particularly desirable when irrigating water is scarce during the dry season. Extra water, when available, will increase the yield, however, and many commercial grape growers have installed irrigation systems.

Grapes are long-lived plants. Once established, they will provide fruit with a minimum of care for a lifetime. Pruning grapevines is an interesting task that can be quite simple or rather complex, depending upon the variety and the form you wish it to take. The most important thing to remember in pruning grapes is that the fresh growth bears the fruit. The European wine grapes often get cut clear back to the stump, leaving only enough dormant buds to start the growth of five or six new canes in the spring. They will have all the grapes that the vine can support for the season.

If enough water and nutrients are available, you can leave an extensive system of old stems and canes to originate a large number of new ones the following spring. Usually it is done on an arbor, trellis, or wire support. I have seen old vines left to grow year after year with no pruning. While they may have still provided a good crop, the grapes were all on the new growth. The old mass of vines did nothing but take up a lot of space and make it difficult to get around it. If you train the canes to a wire or trellis, you can cut them back each year, leaving only one or two buds near the base of the previous year's growth. It allows for better air circulation to prevent mildew and makes it easier to pick the clusters.

Sometimes a late spring frost will destroy the tender new growth after it has grown out a foot or more. It would seem as though the chances of a crop for that year have vanished. Usually, however, some dormant buds haven't grown yet. After the frost, they will replace the frozen canes. The fruit may not be as abundant, and it will ripen a little later than usual, but that is better than none at all.

CHAPTER 6

TREE FRUITS

A fruit tree is a long-term investment that may not yield a quick harvest, but it will eventually offer a reliable source of fruit year after year. If only more people had planted apple trees in their backyards ten years ago, think how many of us would be enjoying the fruit now. Set aside a corner or two for a fruit tree that will appreciate in value each year. Whether or not you reap the harvests, you will add a new source of food to our overloaded ecosystem. A greater number of people planting fruit trees would add much happiness and security to the years ahead.

Since a fruit tree may take up a lot of growing space, you will need to consider how much ground will remain for gardening after the tree matures. If you have only a small garden spot, you may have to forgo the pleasures of a fruit tree and limit yourself to small fruits and vegetables. Don't forget that a full-sized tree will cast a big shadow, and vegetables do best in full sunlight. Even a dwarf fruit tree will interfere if it is close to the garden. Its actively growing root system will literally reach out into the garden and snatch up the greatest share of the available water and nutrients.

Let us assume that you have sufficient ground to plant five full-sized fruit trees. What will they be, and how can you be the most certain of eventually reaping a bountiful crop from them?

Climate will be the first factor to consider. If you live in a cold northern climate, the pome fruits such as apples and pears will be the safest investment. The

Apple varieties in order of ripening from early to late:

July Yellow Transparent   Golden Delicious
    Gravenstein   Red Delicious
    Wealthy   Rome Beauty
    McIntosh   Winesap
    Jonathan   King
    Grimes Golden   *October* Northern Spy

Pollination requirements:

Apples—Most apples are self-unfruitful and need pollen from another variety of apple. Golden Delicious is an excellent pollinator, but any two varieties will pollinate each other.

Pears—Most pears are self-unfruitful and require another pear variety as a pollinator. Duchess d'Angouleme and Kieffer produce quite well without a pollinator. Bartlett and Seckel will not pollinate each other. Clapp's Favorite is a good pollinator for Bartlett.

Peaches—Most peaches do not need a pollinator. J. H. Hale produces better with an Elberta for a pollinator.

Apricots—All varieties will bear without a pollinator but two varieties grown near each other will set heavier crops.

Plums—Japanese varieties such as Santa Rosa need a pollinator. Others such as Stanley are self-fruitful.

Cherries—All sweet cherries need pollinators. Van is a good pollinator that also produces good fruit. Sour or pie cherries do not need pollinators.

Blueberries—Plant two varieties.

Raspberries and Blackberries—All are self-fruitful.

Grapes—All grapes are self-fruitful.

Nut trees—Most need pollinators. Plant two varieties of each kind.

more-moderate-climate fruit trees include the stone
fruits such as peaches, cherries, apricots, and plums.
Warm climates will allow you to grow citrus,
avocados, and figs. If you limit your choices according
to your climate, you will have greater chances of
picking tree-ripened fruits from your own trees. But
we always want to gamble, hoping that maybe we can
find some special variety or trick that will enable us to
beat the weather. Usually we have the frustrating
experience of almost making it but not quite. After
years of painstaking effort, we too often see our
pampered tree killed by a freak storm or bad year. If
not that, we get only enough fruit to tempt us on into
several years of no crop at all, when if we had played it
safe in the first place, we would have a harvest every
year.

Should you have room for only five trees, make at
least four of them sure bets for your climate. Leave one
for experimentation if you enjoy that hobby. It is true
that growers develop new varieties each year. Remember, however, that climates are so variable, especially in the Western states, that you can never be sure
that the highly advertised variety supposed to do well
in your region will actually thrive in your particular
spot.

If you want to make your land an extension of the
state agricultural experiment station for testing new
varieties, then go right ahead, but you should not
plan on much fruit from them. Space for more than
five trees will permit you to try as many experiments
as the size of your land, time, and income will allow.
While the five suitable types will provide a consistent
crop for you, you may come up with something extra
special that you would never have had without trying
a long shot.

Dwarf fruit trees offer a partial solution to the
space problem. In a row capable of supporting five
standard-sized trees, you can plant two rows of ten

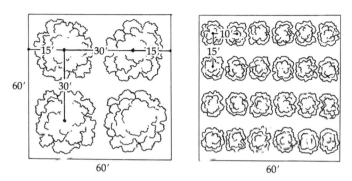

The same space needed for four standard apple
trees will grow twenty-four dwarf apple trees.

dwarf trees or a total of twenty trees. Of course the
dwarf trees will produce a smaller amount of fruit a
tree, but the total crop for the area used will be about
the same. They also allow you to select a greater
number of varieties for the same amount of land. It is
still safer to grow mostly those well adapted to your
climate, but with dwarfs, you can take more chances
and still have a crop even though it may be a smaller
one if you fail.

You can have a tree with fruit on it much sooner
with dwarfs. In fact, many nurseries advertise bare-
root dwarf trees guaranteed to bear fruit the same year
you plant them. How can they be so sure? It so happens that they grew them for perhaps two years under
ideal conditions and care. The trees had established a
healthy root system and set fruit buds before the nursery dug them up. The fat fruit buds develop in late
summer and fall, so it already has the makings of the
coming season's fruit. It is quite a satisfaction to plant
a new tree and pick from it in a few months. But what
about the following year? The nursery only guaranteed fruit for the first season.

A bare-root tree has lost most of its feeder roots in the process of the digging and replanting. The first summer, while the young tree builds a new root system and struggles to establish itself, it doesn't have sufficient reserve to form a new crop of fruit buds. So, after presenting you with fruit the first summer, it most likely will rest the second. Don't think that the nursery has tricked you—they just didn't tell you that your tree would need at least one year to reestablish itself. But, if you have a dwarf tree, it doesn't require several years to reach a balance between root system and top structure so that it can set fruit buds again. Give it plenty of water and keep the weeds away from it, and it may start to bear again the third summer.

It is easier to care for dwarf trees. You don't need a tall ladder to prune, thin, and pick them. When you spray, you can reach the topmost twigs with a hand-pumped sprayer. Since they do not grow as rapidly, a dwarf tree does not need as frequent or severe a pruning. To conserve space and provide decoration, you can espalier them against a fence or fireplace. I have seen dwarf pear trees espaliered against a large brick fireplace almost like a vine. They produce a decorative effect in every season: a mass of bloom in the spring, shiny green foliage in summer, golden fruits and colorful leaves for fall, and a pleasing pattern of branching stems on the red brick in the winter. Where space is really at a premium, a board fence with an eastern or southern exposure can become your fruit orchard in the vertical with little interference to the vegetable garden.

What makes a tree dwarf? Most of them consist of two or three parts of trees grafted together. The grower selects a root from a naturally small variety which never produces an extensive root system. It provides a brake or control on the whole tree since the top depends upon the root system. The stem he may select for strength or pattern of growth, and the top he chooses from the regular varieties of full-sized trees. The fruit will be of the same size, flavor, and texture as that grown on a full-sized tree, but it will be nearer the ground and easier to pick.

Dwarf trees do have a few disadvantages though. Since the root system is not large, the tree may become too heavy, and it may require staking or tying to prevent it from tipping over during severe storms. But in most cases the advantages of dwarfs outweigh the disadvantages when you raise them in small home orchards.

Pruning and thinning fruit trees, either dwarf or full sized, are arts that take time to learn. If you have ever watched a trained horticulturist at work with the pruning saw and shears, you sensed his uncanny ability to anticipate future growth of the tree and thus trim it in just the right way to make it stronger and more productive. You may need to do some reading on the subject and then take a few lessons from someone who has already developed the skill. A few basic principles will help you get started. First remember that the terminal bud or tip of a twig continues the straight-line growth. If you cut the terminal bud off, you will encourage side branching or a more dense growth. When you prune a twig back, the bud closest to the end will grow a side branch in the direction indicated by its position. The first years of pruning establish the trunk of the tree with its pattern of branching. From the start you can set a low crotch or several main trunks by the way you prune the small tree. Each variety of tree has its own pattern of growth and requires a different technique.

A domestic fruit tree properly pruned, fed, and pollinated will usually set more fruit than it can support. Thinning the fruit crop will give you larger and tastier fruit, save the tree from breaking down, avoid extensive propping, and stimulate it to produce every year. It is often difficult to remove enough of the small

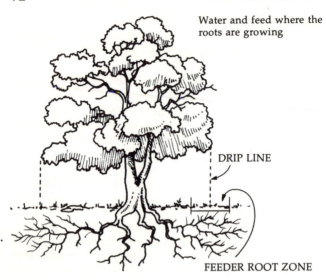

Water and feed where the
roots are growing

DRIP LINE

FEEDER ROOT ZONE

green fruits to achieve the desired results in the mature crop. Here again, experience will help you to know how much you must pull off to have a bountiful harvest of choice ripe fruit later in the season.

Don't forget about nut trees when you plan your home orchard. Most nut trees grow slowly, but like fruit trees they are an excellent investment. Nuts —Mother Nature's storehouse—are high in food value, including proteins. They are easy to store for the winter. Some kinds you can even keep for more than one season. Like other trees, the nut trees have certain climate requirements, but at least one variety exists for almost every climate. English walnuts are not as cold-hardy as black walnuts, but various strains of the Carpathian walnut—similar to the English walnut—are now available that will survive as far north as Canada and withstand 20 degrees F below zero. Vigorous as the Carpathian walnuts are, they may produce a crop in as few as five years.

Each part of the country has its own native nut trees. They offer a good guide as to what cultivated kinds will thrive in a particular area. Wild hazelnuts do particularly well in western Oregon and Washington, as does the closely related domesticated filbert. If you can find wild nuts free for the gathering, you can save your own land for other things, but the domesticated nut trees will usually be more reliable and produce nuts of better quality. Still, if you are fortunate enough to have access to wild pecans, hickory nuts, beechnuts, butternuts, black walnuts, and chestnuts, consider yourself fortunate and take full advantage of it.

Fruit and nut trees are more susceptible to insect damage than small fruits because they live longer and produce their crop slowly over a period of several months. Insects have time to build up large numbers and spread widely. A complex ecosystem is the best approach to pest management of fruit and nut trees, but if you cannot achieve that, you may have to follow a plan called integrated control.

Integrated control is not a program of complete eradication. It combines biological measures with only enough artificial control to hold pest populations to tolerable levels. Fruit you raise for your own use does not have to be perfect in size, or without blemishes. Being likely more concerned about flavor, texture, and ripeness, you can overlook a small amount of insect damage, especially if you produce more than you can eat. Should you need sprays to hold such things as aphids and caterpillars in check, use chemicals readily degradable and employ only enough to achieve the level you need.

CHAPTER 7

LAND—LITTLE OR MUCH

How much land does it take to grow your own food? Many people think it requires at least a five-acre farm. The garden books will tell you that you can grow all the fresh vegetables eaten by a family of four on a plot of ground 15 by 20 feet, or 300 square feet. While that does not include any extra to can or freeze or give away to the neighbors, and it is only vegetables, nonetheless it does illustrate the fact that a small piece of ground can produce a lot of food.

Specific examples have more meaning than generalizations, so let me describe a small country place that produced food for a family of six. My parents were fortunate enough to have two and a half acres on which to raise four children during the depression years. Jobs were scarce and cash almost unavailable, so the only way to guarantee enough food was to grow your own. The fact that we never lacked any is a tribute both to the skill and ambition of my parents and also to the ability of a small piece of ground to raise an abundant crop. The soil was relatively poor, but with proper management, it continued to produce our food year after year.

I can trace in my mind practically every square foot of those two and a half acres. The whole place was about 125 feet wide and 1,000 feet long. Even though we practiced what one could call intensive farming, we never had all of it under constant cultivation. We always had some temporarily unused areas that we allowed to grow up to grass and native vegetation.

A TWO-ACRE HOMESTEAD

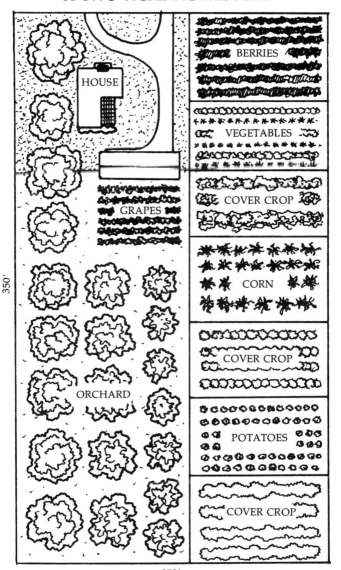

Even two and a half acres provided room for some crop rotation and animal grazing. The vegetable garden itself usually stretched about 75 feet wide and 100 feet long, allowing for generous spaces between the rows for easy cultivation. My parents planted a wide variety of vegetables, including peas, beans of several kinds, beets, carrots, cabbage, cauliflower, spinach, lettuce, chard, corn, squash, pumpkins, cucumbers, onions, parsnips, and sometimes the unusual such as kohlrabi and Brussels sprouts. Tomatoes did not fare as well because of the cool northern climate with abundant rainfall. But we found room for some dill for pickles, kale for the chickens, and huge mangel beets for the cow.

Peas did particularly well. I still have pleasant memories of crawling through the shady tunnel of tall telephone peas on a warm summer afternoon. Picking the long crisp pods, I shelled out handfuls of tender green peas to eat raw until I felt myself about to burst. The peas we planted in a double row about 12 inches apart, leaving room for fence posts and a 6-foot-tall woven wire fence in the center. The vines from both sides climbed to the top of the fence. Often my father would stretch another foot or two of fencing above the 6 feet. Eventually the vines would lop over to rest on the adjacent double row across the 3-foot aisle. The three or four double rows made inviting tunnels to explore.

My family always picked the peas early in the morning, sometimes even before I got out of bed. Early-morning picking meant cool crisp pods, still wet with dew, easy to shell, and not likely to wilt and lose flavor during the process of shelling and canning. Have you ever sat down to shell from a whole washtub filled to overflowing with pods? or maybe even from two or three washtubs? In case you don't know about a washtub, it is about 3 feet in diameter and 18 inches deep. I never could believe that we would get all of those peas shelled, but we all cooperated, and Mother would help while waiting to process the jars in the pressure canner on the wood-burning kitchen stove. Of course we ate all of the peas we wanted raw or freshly cooked with little new potatoes. The rest we canned for winter use, usually in quart jars, often as many as seventy-five quarts. Because we had no freezer, not even a refrigerator, canning was the only way to preserve them.

We let some pods ripen to save for seed for the next year's crop. To keep them free of weevils, which would riddle them full of holes, we used mothballs. I well remember one season when we had an excess of dried peas saved for seed. Mother decided to use some for food, but, alas, even boiling didn't remove the mothball taste.

Potatoes we usually grew in a separate area from the regular vegetable garden and rotated to avoid disease as much as possible. We obtained enough to supply our needs for the winter and have some left for seed next spring. If disease became prevalent, or we desired a new variety, we purchased a small bag of seed potatoes from the feed-and-seed store. Certified seed increased the odds of a good crop. Each seed potato we cut into several golf-ball-sized parts with one or two eyes on each chunk. Sometimes to avoid rot we treated them with a fungicide and allowed them to dry slightly before planting. We buried the seed potatoes in rows about 30 inches apart with hills every 18 inches. One piece went in each hill.

In soil well fertilized with barnyard manure, we simply opened the ground by inserting the shovel and thrusting the handle forward to make a space big enough to accept the potato. Then we pushed dirt in on top with the foot and held it down tightly with the toe while we slipped the shovel out. If the soil was too poor to produce a good crop, we dug a hole about 6 inches deep for each hill with either the shovel or a

hoe and dropped a small handful of commercial fertilizer, probably sulfate of ammonia, in the bottom.
Covering it with about 2 inches of soil, we placed the
seed potato in next and firmed about 3 inches of soil
into place by a light step of the foot on the top of the
hill.

Usually about fifteen to twenty rows 100 feet long
would produce our winter supply of potatoes, which
we stored in the root cellar either in a bin or in burlap
bags. We periodically sorted the potatoes to get rid of
spoiling ones that would ruin the whole stack if not
thrown out. Harvesttime usually came in September
during a pause in the fall rains that would allow for
digging and drying. My father generally dug the
potatoes with a six-tined fork. By inserting the fork
under the hill and lifting with a slight shaking, the
potatoes would rattle free of the dirt. By digging two
rows at a time this way, he saved some extra motions.

I often had the job of gathering the potatoes into
bags to transport them to the root cellar. Eight or ten
sacks of potatoes would suffice. One year my older
brother decided to venture upon a small farming operation of his own and planted a second patch of
potatoes. He hoed and harvested them by himself and
sold them for enough cash to buy himself a new suit. It
was quite an accomplishment for a fifteen-year-old
boy, especially since potatoes went for only a dollar a
hundredweight sack.

Now, remember, we are still talking about only
two and a half acres of ground, and I have much more
to tell. Since carrots would also keep through most of
the winter in the root cellar, we always had several
rows in the vegetable garden. I can remember a few
times when I had to weed a row of carrots before I
could go swimming in the lake, and the rows seemed
interminably long. Years later on a return visit to the
old homeplace, I saw how short the rows really were
and how small the whole garden plot seemed in con-

trast to my childhood concept of the whole world on two acres. The incentive to weed the carrots involved more than just a chance to go swimming. Early in the season I started harvesting those carrots. When they reached the size of a lead pencil and 4 to 6 inches long, I would pull a small handful on my way to the lake. There I would wash the dirt off and enjoy their crisp sweetness while dangling my dusty bare feet in the cool water.

Carrots were a common item on our table through the summer and fall. We stored them in the root cellar with the potatoes before the frost got them. Sometimes Mother canned some, too, to eat boiled in stew or as tender slices mixed with green peas. If you thin out the carrots in the row to leave enough space between for expansion, they will grow into nice smooth taproots that you can easily clean and store. Overcrowded, they will be knobby and twisted around each other so that it is difficult to pull them without damaging the outer layer of cells. Any break in the surface will invite rot and decay.

Remember that the carrots you store in the root cellar are not dead. They are alive, and you must not allow them to dry out, or they will shrivel and spoil. Had you not dug them, they would remain alive in the soil through the winter (if protected from frost and rot), to grow again next spring.

During the second year the carrot will blossom and produce seed. If you live in a mild, dry climate you can leave carrots in the ground and dig them only as you want them. However, you must consume them before spring, or they will send up a tall blossom stalk and go to seed. Our root cellar maintained a high humidity and moderate temperatures which kept the carrots alive and in good condition throughout the winter. If we had some left over in the spring, we fed them to the cow before they began to grow or spoil.

The large mangel beets, related to the sugar beets,

also went into the root cellar. Some were 8 inches in diameter and up to 18 inches long. We stored them just for the cow in order to provide some variety from dry hay during the cold winter months. I remember watching my father cut them up with a hatchet or heavy knife. Crisp and clean and white, they looked so good I was tempted at times to sample them. Being starchy, they didn't excite my taste buds, but when I went to the barn in the evening to milk the cow, she seemed to know if I had mangel beets in the bucket with her ration of grain feed. She could hardly stand still until I had placed the bucket in her manger.

The mature mangel beets were only for the cow, but we got our share of them earlier. Since they grew so large, we had to thin them out to several inches apart in the row while still small. When the tops stood about 6 inches tall and the beets were only as big as marbles, they were much superior to red beets for beet greens, at least in my estimation. Tender and succulent, they had a much sweeter taste than common red table beets. Mother even canned a few. We had plenty of red beets, too, and we also thinned them for greens. Mature red beets would keep for a while in the root cellar, but in addition we canned some for winter use.

Beans we ate in season and canned in two forms. The string beans or green beans ripened first. We had to pick them just at the right time to avoid unchewable strings. Besides them, we grew cranberry beans (now listed in the seed catalogs by the uninviting name of horticultural bean). As I mentioned earlier, you could use them as green beans while still immature. But when they became fully mature, but not dry, you could shell out the big, fat, meaty beans. We called them simply shell beans or shelled beans. One could use other varieties for that purpose, but we preferred the cranberry beans for shelling. Their flavor was such that they didn't require catsup, molasses, sugar, or spices. One could simply boil them in water with a little salt and butter. Mother usually canned about thirty quarts of them, frequently mixing them with whole kernel corn to make succotash. Some pods we left for the beans to ripen and dry. Of those we kept a portion for seed, and the rest we boiled like pinto or navy beans.

One other bean we often grew on our place, but it was not for food—only for fun. We planted scarlet runner beans—later covered with bright scarlet blossoms and dark green leaves—next to a fence or wall where they could climb six to eight feet high. The big pods, when dry, held inch-long beans of a rich purple hue, slightly mottled with gray and shiny as if varnished. They are not edible, at least not when we had others so much better. But they added a decorative touch that we enjoyed.

Recently my wife and I grew some in our garden just for old time's sake. I shelled out a pint of them, and the rich purple color (purple is her favorite) so intrigued my wife that she put them in a decorative wooden bowl on the side table where we can enjoy looking at them. Now and then we scoop them up and let them fall through our fingers. Don't forget that vegetables can feed the soul as well as the stomach.

Lettuce held a prominent place in the garden on my parents' two and a half acres. Leaf lettuce, usually the bronze-edged curly leaves, was one of the first fresh vegetables ready to eat in the spring—a welcome change from the canned and cooked vegetables of wintertime. A large bowl of freshly picked and washed lettuce leaves on the table provided a salad or, placed in the fold of a slice of homemade bread and butter, a crunchy sandwich. Head lettuce came a little later when the leaf lettuce began to turn bitter.

Cabbage heads began to form about midsummer. By planting both early- and late-season varieties, we could have cole slaw over a period of two months. I can remember some "before dinner" salads that I had with a generous section of a cabbage head in one hand and a salt shaker in the other while I leisurely wandered through the flower garden observing the butterflies on the lilies or sniffing the roses.

The late cabbages we left growing as long as possible. They frequently produced mammoth heads in the late summer and early fall. Before the hard frosts began to soften and rot them, we shredded and packed them into a large earthenware crock with salt to make sauerkraut. By the time the sauerkraut was ready to eat, we had begun to empty some of the 650 quart jars of fruits and vegetables we had canned during the summer and fall. The empty jars we then filled with sauerkraut and processed and sealed them for later use.

A few cabbages we stored in the root cellar with the carrots, potatoes, and onions to use in making stew or a boiled vegetable dinner. Brussels sprouts would sometimes keep on producing little miniature cabbages right into the winter season to provide us with fresh green vegetables after all other green things had given up.

The corn ripened in late summer, and we feasted on roasting ears while we could. The eight to ten rows of sweet corn provided a good supply. Mother usually canned thirty to sixty pints of corn. A little more mature than the freshly boiled roasting ears, we blanched and cut it from the cobs. It was full of flavor and starchy sweetness. Some of the choice ears we allowed to ripen completely. When dry, we shelled and stored them for seed for the next year, or we parched and ate the corn during a long winter evening. I think we kids ate parched corn mostly for the butter and salt. Now I marvel that we didn't break our teeth on the hard kernels that puffed up round and smooth in the hot greased frying pan. Watched carefully and stirred frequently, they had a pleasant roasted corn flavor. If the fire got too hot or we did not stir them enough, they tasted more like oiled charcoal than anything else. But it was fun anyway —something different.

Much better than parched corn was the fluffy, flavorful popcorn that we enjoyed frequently during the winter. We grew our own in a patch several hundred feet from the sweet corn to prevent cross pollination. One year I planted extra and sold the surplus to neighbors for a few pennies. The popcorn matured rather late and had to fully ripen and become fairly hard before being picked. Ring-necked pheasants discovered my popcorn patch a week or two before it was ready to harvest, and they decided it was a delightfully easy place to fatten up for the winter. They would jump up, grab the husks with the beak, and pull them down to expose the kernels. Then they would flutter up and peck a kernel at a time until if lucky they would knock the whole ear to the ground. Should that happen, they would quickly strip it clean. I know all about it because I caught them in the act almost every evening after school.

As soon as the school bus stopped to let me off at the driveway, I would dash into the house, throw down my books and lunch box, and race to the corn patch to chase the pheasants out. I think they soon learned my schedule and saw to it that they had their crops full for the day before I got home from school. At any rate, even after sharing with the pheasants, I had about as many ears as I had time to pick and hang up by the husks to finish drying.

I spent several rainy winter evenings shelling the hard kernels from the dry cobs. By rubbing one ear against another, I could dislodge the tightly packed kernels. It became a sort of game to see which ear was the toughest and would shell off all of the others. Then came the winnowing on a windy day. The mixture of corn and chaff I poured from one large container into another, with the wind blowing away all of the light particles of cob and other debris. It gave me a great deal of satisfaction to lift the sack of twenty pounds of popcorn. I measured out about ten pounds for our own use and sold the rest, if I could find someone who had money to buy it.

One year I grew a Japanese hull-less variety that was supposedly superior in several ways. Every kernel was pointed like a grain of rice, and it had a sharp curved spine at the tip. After shredding a good deal of hide from my hands while handling it, I decided the good old American variety with smooth, round, yellow kernels was more to my liking.

Popcorn was the main feature of many wintry Saturday nights at home by the roaring fire. Neighbors might join us for an evening of simple games like dominoes or Parcheesi, and all helped themselves to bowlful after bowlful of freshly popped corn liberally buttered and lightly salted. The corn we popped in a screen-wire basket on the end of a long wire handle, not over the coals in the fireplace, but on top of the old wood-burning kitchen range. If anyone was cold, a few turns at shaking the popper over the hot stove would warm him in a hurry. We made it by the dishpanful, and everyone had all he wanted. I particularly remember the refreshment of a glass of cold milk from the tall pitcher in the cooler after many bowls of dry, salty popcorn.

I should say something about tomatoes in the vegetable garden, but not from my childhood experience. The cool wet summers in our climate made tomato growing a most unsatisfying venture. About the first of March, Mother always planted some tomato seeds in a flowerpot she placed near a south-facing window. The seedlings grew nicely, and we set them out in the garden around the first of May. The plants grew slowly through the cool days of May and June but began to prosper in July and August.

However, the shorter, cooler days of September slowed them down. Before the frosty nights of late September or early October spoiled them, we picked the largest of the tomatoes, still green as grass, and carefully wrapped each one in paper and packed it into a box that we kept in the house. Every few days we inspected the tomatoes, threw any rotting ones out, and placed the most promising ones on the kitchen windowsill to ripen. They did turn red, and we ate them, but not with the same pleasure as we did the dark-red ripe tomatoes shipped in from warmer climates.

Sometimes we sliced and fried the green tomatoes. Green or ripe, they made good sweet preserves if you added enough sugar to practically candy them. Some years we could afford to buy a few crates of tomatoes from the store to enjoy fresh and also canned for the winter.

Fruit was an important part of our diet. When canned, colorful rows of the various fruits available to us lined the shelves in the root cellar. We always had a generous-sized strawberry patch on our two and a half acres. It required a considerable amount of handwork, but the reward was well worth it. During the winter season a liberal application of barnyard manure or chicken litter provided the essential fertilizer for vigorous plants and choice berries.

Several varieties did well in our area, but I particularly remember the Marshalls. The plants had large, rich-green leaves, and the berries were big and juicy. The birds helped themselves to their share, but the eight rows about 75 feet long provided plenty for all. Some of our extra we gave to friends and neighbors, since no one had money to buy them. One summer we traded many flats—I cannot remember how many—to the photographer in town for a picture of my baby sister.

In season we had mashed strawberries for breakfast, strawberry shortcake for dinner, and a bowl of strawberries with our supper. It seems as if we never grew tired of strawberries. Even today I enjoy strawberries.

Have you ever eaten canned strawberries? Don't bother if you haven't. They are rather disappointing after eating fresh ones. We had no freezer (not even a refrigerator), so with such an abundance of fresh strawberries, we attempted to preserve some of them for winter. Mother usually canned fifty to seventy-five quarts by the cold-pack method. They looked beautiful in the jars just before she put them into the boiler for processing. When they came out, all of the berries had collected into a tight mass in the upper one fourth of the jar, right under the lid. The bright red juice filled the rest of the jar, leaving the berries a dull pink color. When we opened the jars during the winter, we separated the mass of compressed berries as best we could and spooned them up with a liberal portion of juice. We enjoyed them, even if it was a far cry from strawberry shortcake. The flavor, watered down, was still there, and it reminded us of warm, sunny summer days.

Red raspberries grew nicely in our climate and required relatively little care. Three rows, the same length as the strawberry patch, gave us all we could use. We rarely cultivated the raspberry patch. I can remember wading through knee-high grass between the rows to pick the large berries. They might have produced a heavier crop if trained on wires and cultivated clean, but we were not growing them to sell and had other more urgent things to do. Every year we canned fifty to seventy-five quarts, and they turned out somewhat better than the strawberries. Still not as good as frozen raspberries but full of rich flavor and nutrients, they helped to balance our winter diet.

Black raspberries grew wild in the woods. We transplanted a clump of them into a corner by the chicken house where we could care for them. While more seedy than the red raspberries, they had a distinct flavor of their own that added particularly to the berry jams that we enjoyed with our meals, especially in our lunches at school.

Wild blackberries, a scourge of the countryside, grew into inpenetrable thickets on every fencerow and at the edge of every clearing in the woods. The berries were large and juicy, quite sweet if fully ripe, but with relatively little flavor. Literally tons of them went to waste every year, but we canned some. Frequently, though, they got pushed to the back of the shelves, other fruits being preferred before them. Although a bit too seedy for jam, they made lovely jelly.

One species of wild blackberry was extremely delicious in flavor but difficult to obtain. We called it the running blackberry, something like the dewberry of other regions. This kind grew almost everywhere through the woods, slender trailing vines on the forest floor that seemed to grab your toes and trip you at every other step. But it almost never produced any fruit except where someone had slashed the trees to form a sunny opening.

In recently logged areas with their massive tangles of broken limbs, stumps, and fallen trees, the running blackberries would grow in the full sunlight and bear a good but scattered crop. To get them meant scrambling over, through, and under the wreckage of the previous forest. The thorns and the hot sun did their part to discourage us, and the berries were small, filling the bucket with agonizing slowness. It took only a large patch of blighted blossoms or a nest of mean hornets to send us to a convenient huckleberry bush or a clump of black raspberries to rapidly finish filling our buckets, and so we never seemed to manage to get as many running blackberries as we really wanted. Maybe that's why they tasted so extra good.

Blueberries grew quite well in our area, but unless you had a large patch, the birds ate them all before they got fully ripe. Half-ripe blueberries aren't really worth the work it takes to grow them. We solved that problem by turning to the woods again. In the higher elevations near timberline, a low-bush blueberry, *Vaccinium deliciosum*, produced a crop true to its name. Some years they did not ripen well, but in good years we often went there to get our share along with the black bears and chipmunks.

Tree fruits did not form a large part of our family farm, but we did have several varieties available to us. My parents' property contained two apple trees, one Gravenstein and one crab apple. We children started eating apples as soon as they began to show color, or even before. At harvesttime we carefully picked and stored the fruit in boxes for later use. Apples would keep until midwinter when we had some emptied canning jars to refill with applesauce. If our tree did not produce enough, apples were going to waste in small orchards scattered throughout the community. We could usually obtain them just for the picking. Likewise, we could harvest cherries, both sweet and sour, either free or for a small price, or maybe in trade for some item that we had in extra supply. Since pears and prunes were often similarly available, our shelves usually contained four to five hundred quarts of canned fruit for our winter supply.

Our two and a half acres also provided forage to support one cow through the winter. The lower acre we would periodically seed to a hay crop. As much barnyard fertilizer as we could spare went onto it. We never had enough manure to satisfy the needs of the whole two and a half acres, but by alternating from one plot of ground to another, we maintained the fertility sufficiently to produce an average crop of hay. Our cow grazed the hayfield through the winter, and then during late spring and summer we pastured her on the neighbors' land where native grasses and herbs grew at random over several hundred acres, enough to support several head of cattle. We usually made some kind of exchange for the grazing privilege—sometimes extra milk and butter or a certain number of days of work on a project that required extra help.

The crop of hay we mowed by hand scythe in the late summer when we could expect enough days without rain to cure it and get it into the hayloft. Generally it involved the use of a neighbor's team and wagon, which we also paid for in barter or labor. Managing and maintaining the milch cow on two and a half acres were probably the most difficult parts of our whole food producing procedure, but the milk, cheese, and butter constituted an important part of our diet—especially for our protein needs.

We always had some chickens. For several years we enlarged our flock to several hundred in an attempt to bring in some cash. However, the margin was so narrow in a poultry operation where you had to purchase all of the feed that we never realized much more than a lot of hard work, plenty of eggs to eat, an increased amount of fertilizer, and sometimes an unpaid chicken-feed bill. Only the large producers made a living at it, and during some years they ended up with more outgo than income. Our land could have easily housed a small flock of fifteen to twenty chickens with enough extra eggs to trade at the store for feed other than what we could produce.

The main conclusion I would like to draw from my description of my parents' two-and-a-half-acre farm is this: It is possible to raise enough fruits and vegetables for a year's supply for a family of six on one acre of ground. One can add eggs with a little more ground if he can purchase some chicken feed, especially during the winter months. Milk for the family would require more land suitable for pasture and hay. The actual amount would vary according to climate and soil, but three acres if managed and maintained properly could provide forage and hay for one cow. An alfalfa region requires much less acreage than if you can grow only light crops of timothy or wild hay.

So, what more do you need? A small patch of wheat will give you flour, and a few swarms of bees, your sweetening. Nut trees can supplement your protein. But the country way is more than just food in your stomach. You need a comfortable home and clothing, some shelter for animals you might keep, and tools with which to do your work. And don't be so naive as to think that you can do much living in the present world without some cash income, even on the barest subsistence level.

Where will you find land for your own country-living experience?

Land for sale exists in all parts of the country, but which area is best? I hope you will heed the advice contained in the many books written about purchasing land and investigate thoroughly before you start payments.

What should you look for?

First of all, what part of the country would you like to live in? Since you have considerable choice in the matter, why not select a region in which you would be most comfortable? What type of climate do you prefer? Most people select the setting in which they spent their most pleasant years. For many, those are the years of childhood when they had no responsibility for making a living or feeding a family. Be careful. While you enjoyed sledding with your friends on the snow-covered hill, someone else paid the fuel bill to warm the house. As you splashed in the cool water to temper the heat of the desert sun, another worried about how to get enough water to keep the garden alive. And while you sat in the schoolroom watching the raindrops trickle down the windowpane, a farmer tried to get crops in between showers. The perspective changes when you become responsible for making a living.

On the other hand, you may be more willing to put up with inconvenience and hardship in a familiar climate and setting than one that seems foreign. I have never been able to fully adjust to a dry climate, because I learned as a youngster to live in a wet one. Some of my friends cannot understand why I would ever choose to live where it rains so much. Sure, I know about the inconveniences of the rain, but I also know of its advantages and how to adjust to it. On the other hand, I wonder why some of my friends would decide on a dry climate where they have to irrigate everything to make it grow. No perfect climate exists anywhere on earth, and it is probably a good thing it doesn't, or everyone would be fighting for a place to live there.

Certainly, if you could live for a full year in an area before deciding to purchase land there, you would have a much better basis for your decision. But that is not always possible. You can read statistics from the chamber of commerce on the weather of the locality, but even that can be misleading. For instance, the average annual rainfall for large areas of northern California is 40 to 60 inches. That is the same as for parts of northwestern Washington State, but what a difference! The rain in northern California all falls during the winter months, and the four months of hot summer pass without a drop. On the other hand, the rainfall in northwestern Washington spreads throughout the year in a sometimes monotonous drizzle that seems to never stop.

What is the best area in which to grow your own food? If you could spend time in travel and research, you would find that someone has learned to raise it in almost every part of our country except in the high mountains. The kind of crops may differ, but if you adapt your diet, you can survive just about anywhere. But here again, you will probably be most satisfied in a climate that produces the kinds of foods you like most.

Is there any place where you can still buy cheap land? Not really. Everyone now realizes that many people are looking for a country place, and he is not about to sell for anything less than the current market prices. If you have spent some time looking for bargain-priced land, you have no doubt discovered that the low-priced land is not good for anything. Some reason always exists why the seller wants to get rid of it so cheaply. It may flood in the winter, or the soil may be rocky and poor. The land might have no access roads or be of disputed ownership. You should save money in the long run by buying through a reputable real estate agent, even with the commission he must charge. If you find a good deal from a private individual, have the contract of sale written up by an attorney and insist on title insurance.

Usually you will need a larger down payment on country property than on a subdivision tract, often as much as one third of the purchase price. Farmland with buildings on it may be less an acre after subtracting the value of the buildings than just bare land. Don't fall for the idea that you can fix up an old house at less cost than you can construct a new one. With the prices of material and labor as they are now, it often costs as much to rebuild an old house as it does to put up a new one. Even if the cost is less and you can do most of the work yourself, you still have an old house when you get through that is not as valuable or salable as a new one. Should you not have sufficient capital funds to invest in a good house but are willing to live in an old one for a long time while you fix it yourself, then give it a try, though at least be realistic and don't expect things to improve quickly.

CHAPTER 8

WILDERNESS LIVING

Probably everyone spends some time dreaming about a pioneer experience where he can meet nature head on and prove his ability to solve life's problems without all of the complexities of our so-called civilized world. Now, how far backward from where you are do you need to go to satisfy the desire? You have probably seen, as I have, advertisements in travel magazines designed to attract people to a certain type of wilderness experience. A beautiful picture of Lake Louise in the Canadian Rockies appeared above a description enticing the prospective adventurer to a vast panorama of wild beauty that he could enjoy from a comfortable heated hotel room and with access to a heated pool, game rooms, elegant dining and dancing. The copy ended with "Ah, Wilderness!"

My wife and I have always received great benefit when we backpacked into one of the national forests. There we could find the unhurried pace that brings peace. For just a few days we could reassure ourselves that it is possible to live in harmony with nature. With nothing more than physical fitness, mental alertness, and a few essentials, we could live happily and comfortably for a week so far removed from "civilization" that it need not really exist.

In recent years we have discovered another type of wilderness living that does not necessarily replace the backpacking expedition but can add more of the same sense of well-being and satisfaction. After a diligent search we found a small piece of land that fits our

budget and life-style as a future home in the country.
It certainly doesn't match our dreams of a secluded
valley at the foot of towering mountains with water-
falls, big trees, a spring-fed creek, and a lake at the
foot of well-drained, fertile acres. But it does have
something else that seems more important than any of
the above. It has our name on it, which gives us full
control, within reasonable limits, as to what it shall
eventually become. And it is densely wooded enough
and situated in such a way that we can camp in the
middle of it without seeing anyone around us. At the
moment that seems desirable; later on it may lose its
significance.

At any rate we can face the challenge of the wilderness on our own piece of land and do whatever we
want to in whatever way we might decide. To my way
of thinking that satisfies the basic desire for a pioneering experience more completely than just surviving in
the wilderness for a week. We can use our ingenuity
and creative abilities to live in as primitive a setting as
we might like, or we can modify the land to provide us
with an experience of some other pattern.

Some of my friends have found enjoyment from a
much smaller and less isolated piece of ground than
mine. And I know one individual who obtains his
pioneering experience on less than one acre of ground
by growing his own food and building his own
equipment with the least possible dependence upon
modern technology. The fact that he has neighbors on
all sides and is only a few minutes away from a shop-
ping center does not lessen the enjoyment he derives
from doing things for himself. If you approach it with
the proper attitude, living on the land can demand as
much skill and ingenuity today as it did two hundred
years ago.

I suppose the desire to build a house is instinctive.
Like two little savages, my cousin and I started in-
numerable log cabins in the woods. We usually got as

far as a 20-foot square cleared of brush and four poles notched and properly laid as a foundation before time to go home for supper. By the next playday we had either forgotten the location or found a much better site on which to start all over again. The desire stays with an individual throughout his lifetime and often gets expressed through designing or constructing a modern home.

Country living will involve either building a house or improving an existing structure to suit your likes and needs. Should you have the opportunity to start one from scratch, you can gain a great deal of satisfaction from planning and building it as much as possible by yourself. In some cases it may be more of a necessity than a choice. With others it may be a matter of deciding what you like to do most and delegating the least interesting work to someone else.

Not many places still remain where you can buy a piece of land with the proper materials on it for building a log cabin. Don't let that stop you. Buy the next closest thing to it and erect your own anyway. Milled boards and timbers can form as simple and functional a structure as your imagination can design. Or consult one of the many books on rustic house construction and adapt an existing plan to fit your needs. If you do have a place with enough trees for a log cabin, you can obtain excellent books on how to fashion them into a home.

Shortages of fuel and electricity have intensified the incentives to make a country place simple and uncluttered with modern gadgets. A home that you can keep comfortable without electricity or petroleum products may be a real refuge in the years ahead. Wood-burning stoves and fireplaces are becoming more than decorations again. Unfortunately most people have either forgotten or never learned how to build a proper fire in a cookstove and to maintain a steady heat for cooking and baking.

The U.S. Forest Service has recently published an excellent little pamphlet on selection and use of firewood. The following table from it may help you decide on what kind to obtain for your wilderness home.

It computes the efficiencies of various kinds of wood as compared to oil as fuel. It is interesting to know that at current prices most wood is a better buy than fuel oil for heating your house. The figures vary of course with the efficiency of your wood-burning stove (the best are around 50 percent compared to 60 percent for an oil furnace) as well as the dryness and variety of the wood you burn. As fuel shortages increase and people learn about the value of wood as fuel, the price for a cord will also increase. A few acres of productive woodlot may become more valuable than you once thought. In addition to supplying your own needs, it may if managed properly provide some cash income for you. Such facts have some far-reaching implications.

Will we rapidly deplete our wood resources as well as coal and oil? One factor in our favor is the renewable nature of our forests. While we could warp the statistics from the lumber industries and the U.S. Forest Service into all kinds of meanings, it does appear that the eastern hardwood forests are growing wood faster than we cut it—partly because the large commercially valuable virgin maple, beech, and oak are all gone and no one is lumbering the smaller second-growth forests. Proper management could produce large quantities of firewood and at the same time increase the amount of commercially valuable hardwood species for lumber.

Transportation of the firewood would, of course, limit its replacement of oil. It seems unlikely that anyone will suddenly exploit our forest land to relieve the energy crisis. However, since small private landowners have most of the forest land, it is comforting to know that they will have fuel in spite of the oil short-

RATINGS FOR FIREWOOD

	Relative amount of heat	Easy to burn	Easy to split	Does it have heavy smoke?	Does it pop or throw sparks?	General rating and remarks
HARDWOOD TREES						
Ash, red oak, white oak, beech, birch, hickory, hard maple, pecan, dogwood	High	Yes	Yes	No	No	Excellent
Soft maple, cherry, walnut	Medium	Yes	Yes	No	No	Good
Elm, sycamore, gum	Medium	Medium	No	Medium	No	Fair—contains too much water when green
Aspen, basswood, cottonwood, yellow poplar	Low	Yes	Yes	Medium	No	Fair—but good for kindling
SOFTWOOD TREES						
Southern yellow pine, Douglas fir	High	Yes	Yes	Yes	No	Good—but smoky
Cypress, redwood	Medium	Medium	Yes	Medium	No	Fair
White cedar, western red cedar, eastern red cedar	Medium	Yes	Yes	Medium	Yes	Good—excellent for kindling
Eastern white pine, western white pine, sugar pine, ponderosa pine, true firs	Low	Medium	Yes	Medium	No	Fair—good kindling
Tamarack, larch	Medium	Yes	Yes	Medium	Yes	Fair
Spruce	Low	Yes	Yes	Medium	Yes	Poor—but good for kindling

See USDA leaflet No. 559, "Firewood for Your Fireplace—Selection, Purchase, Use."

ages. If you have such a woodlot, be sure to learn how to take care of it properly so you don't squander your advantage.

Supposedly, you can heat the average American home with five cords of firewood a year. Another estimate tells us that in a year the average woodlot produces one cord of usable firewood for each acre. Thus you will need about five acres of woodlot to provide a continuous supply of fuel. If you live in a warmer-than-average climate, you will get by on less. In a colder one, you will need more. The species of tree will make some difference, too, as well as your ability to protect your forest from disease and fire damage.

The people of Europe have learned after many generations that forests are not indestructible, and if misused they may disappear forever. We should learn from their experience, but mostly we go on pasturing livestock in our woods to eat up or trample down every little seedling tree. Also we cut large areas clean without planting a single tree to replace them. When we do replant, it is usually all one species, which produces a monoculture system with all of its problems. While we blame big industry and the Forest Service for destroying our forest resources, in reality it is the small landowner who refuses to learn how to grow trees. Tax dollars pay for an abundance of information in free bulletins and a vast corps of public servants to offer free advice, but too few people take advantage of them.

Today we hear a great amount of discussion about using solar energy. But have you ever considered the fact that our forests are already solar converters? Photosynthesis ties up the light energy from the sun in the wood fibers. Burning fossil fuels such as coal and oil releases solar energy trapped by the plants and animals living on the earth before us. Combustion releases large amounts of carbon dioxide. Carbon dioxide combines with water in the presence of chlorophyll and sunlight to produce carbohydrates and oxygen. The trees reduce the carbon-dioxide level and increase oxygen, so essential to our existence.

What about the smoke from many wood-burning stoves? Will the increase of wood for fuel lead to greater pollution problems than we already have? Where we live no one can burn anything out-of-doors. Leaves, trash, and paper we must place in the garbage to be hauled to the county dump. It is supposed to reduce air pollution. If we have time and room, it is perfectly all right to pile the leaves up in a heap and let them rot. What, then, is the difference between incinerating the leaves in a bonfire or letting them decay in a heap?

To answer that we must know what burning is. Leaves and branches heated to their kindling point burst into flame and are consumed except for a little ash and smoke. The smoke contains some unburned carbon—the visible part—and various oxides of carbon such as carbon monoxide and carbon dioxide plus a variety of other volatile substances in small amounts.

The rotting trash heap is also oxidation but at lower temperatures. It has no flame, and the bacteria provide the enzyme systems to convert the fuel to heat, carbon monoxide, carbon dioxide, volatile gases, and ash. In fact, if you do not turn the mass, the oxidation process may generate so much heat it will turn to white ashes similar to those from the rapid combustion of a fire.

The two processes start with the same raw materials and produce similar products. In the case of the fire, some partly burned hydrocarbons go into the air as smoke along with the carbon dioxide. It is the more harmless type of air pollution that will settle out rather quickly. The rotting heap also gives off carbon dioxide in large quantities along with other volatile substances, but the partly burned hydrocarbons remain on the ground. The carbon monoxide from either process is relatively insignificant and readily breaks down in the natural environment. When you consider the entire ecosystem, burning wood in the stove doesn't add or subtract anything more than rotting leaves in the forest.

Fossil fuels, however, such as coal and oil release sulfur compounds into the air. It is the sulfur compounds such as sulfur dioxide that comprise the more dangerous parts of air pollution. The pollution from burning wood is therefore not as serious as that of burning coal and oil. The natural process of decay will return the elements of the wood back to the ecosystem anyway. Using wood as fuel frees the solar energy in a form that can bring us comfort.

The woodlot is a good type of wood storage. What you do not cut will not only keep but continue to increase in volume. You will never get too much wood. If you let some of the trees mature, you can sell them later for lumber. And should you saw more firewood than you need for the year, it will last until the following winter. In fact, most wood needs to dry for several months, and if preserved properly, it will make just as good fuel several years later. For long periods keep the wood dry. You can stack well-seasoned wood closely in a woodshed or in a covered stack. Until it thoroughly dries, however, you must pile it loosely so air can circulate through the stack. Split any wood chopped during the growing season, when the tree is full of sap, into pieces 4 inches or less in diameter to let it dry faster and discourage rotting.

How much wood is in a cord? Most correctly the term applies to a stack of wood 4 feet wide, 4 feet high, and 8 feet long, or with a total volume of 128 cubic feet. That includes the air between the pieces so it is not as exact a measurement as board feet used in measuring lumber. Since most firewood does not come in 4-foot lengths but rather 16 inches long or less, a full cord

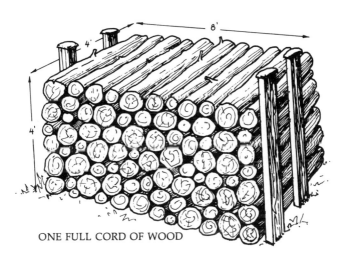

ONE FULL CORD OF WOOD

will require at least three stacks 4 feet high and 8 feet long. Various terms exist for one of the stacks. In some places people call the single stack of 16-inch wood a cord, but that is really a misnomer—it is only one third of a cord. Others term it a "face cord," which would apply to a stack 4 feet high and 8 feet long, no matter how long the pieces were. A stovewood cord might mean a stack of 16-inch pieces 4 feet high and 8 feet long while a fireplace cord would be a stack of 24-inch pieces 4 feet high and 8 feet long. In some areas a single stack goes by the name of rick. Three ricks of stovewood 16 inches by 4 feet by 8 feet constitute a full cord, as do two ricks of fireplace wood 24 inches by 4 feet by 8 feet.

Most hardwoods of high density will weigh close to two tons a cord even when dry. Don't expect to haul a cord of wood on a pickup truck unless it is very dry softwood of lower density. The amount of fuel energy in a cord of hardwood is much greater than that in a cord of softwood. For instance, oak will produce more heat energy than soft pine. If your tools are sharp, it doesn't take any more work to cut a cord of oak than it does a cord of pine; so, if you have a choice, select the oak. When buying wood, obviously you pay a higher price for oak or hickory, but you get your money's worth in heat.

Sometimes completely dried wood is not really the best. Wood with a low moisture content will burn rapidly and release its energy all at once. While that is good for kindling to start a fire with, once it is going well, wood with some moisture content will hold the fire longer by burning more slowly. The semidry wood puts out almost as much heat, but it will do so more slowly. Wood too wet to ignite easily you must add to an already hot fire or split into small pieces so the flames will dry it out quickly and start it burning. A mixture of dry and semidry wood may best maintain a steady fire; especially is this true of the less-dense woods.

Kindling wood is important for starting fires. Split it into fine splinters that ignite rapidly so as to generate heat sufficient to start larger pieces. Most people use a few sheets of newspaper to light the kindling. But if you do not have any, you can cut some of the kindling into splinters to set afire with a match, then add progressively larger pieces. In the old homestead days it was a common practice to shave some of the kindling into "fuzz sticks" that took the place of paper. The fuzz sticks lit easily with a match, but then the pioneer needed some small kindling handy to lay carefully on them. Larger kindling and small sticks of firewood followed until the fire burned steadily with a good bed of coals. The pioneers often prepared their fuzz sticks just before retiring. With a good supply of kindling and dry fuel handy, they could have a roaring fire chasing the chill from the room within a few minutes after getting out of bed.

Dry cedar or spruce makes the best kindling. The straight-grained western red cedar used to make shingles and shakes is excellent. Such woods are becoming increasingly difficult to find, however, so you may not enjoy such luxury. Kindling made from large pieces of wood must be straight grained and free of knots. You can employ almost any kind of wood if you can split it into fine enough pieces. Small branches and prunings from yard trees, you can stack and dry for kindling later.

The tools for cutting wood will vary according to the nature of the trees and how much you need. Man hewed many thousands of cords of wood before the powered chain saw ever existed. Some people still do not have a power saw, or for various reasons they prefer to use hand tools for the job. If time is not a factor, you may find a great amount of satisfaction and pleasure in using simple tools to convert a tree into fuel for the stove. Once you understand the tools and know how to keep them in good condition, the task can be a pleasant experience and excellent exercise. However, a power saw can allow you to spend more of your time on other chores.

The ax is the first essential for any woodcutting operation. The kind of ax is mostly a matter of preference. Usually it takes a little experience to determine the weight and style most efficient for you. It should be heavy enough to carry through a swing with good force but not so much that it wears you out lifting it into the air. The amount of time you spend working with an ax will determine the muscle that you develop. As your strength increases, you can handle a bigger ax and gain the advantage of the force of a heavy moving object. To state it another way, you must apply more muscular force on the downward swing of a light ax to make the same cut. On the other hand it may be easier to learn to hit the same spot twice in a row with a light ax because you can control

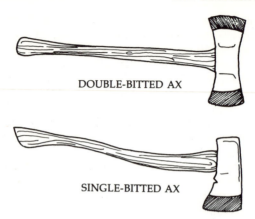

DOUBLE-BITTED AX

SINGLE-BITTED AX

it better. Many woodsmen prefer a double-bitted ax, claiming that it has better balance than a single. Whatever you select, be sure to keep it sharp. Dull tools require great expenditures of energy for little work done. Sharpen your ax with a flat mill file run against the cutting edge. A file guard may save you from scarred knuckles.

Different handsaws have specific purposes. Before the powered chain saws, the woodsman's crosscut

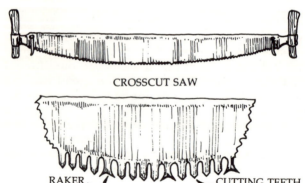

CROSSCUT SAW

RAKER   CUTTING TEETH

saw did the job. For general use a 6-foot blade with handles at both ends is suitable. The saw is heavy enough so that you need only to draw it back and forth over the log. Here again, sharpness is the key. Saw filers do not learn their skill in a day or a year. It is generally the older person who has spent many years both using and sharpening saws that has developed the keen touch required to make a crosscut saw bring out 3-inch shavings with cat whiskers on the end. When you think about all of the giant trees from the redwood region and the fir forests of the Pacific Northwest that men harvested with hand tools, you realize that they were designed and maintained for the greatest efficiency possible.

BOW SAW

SAWBUCK

Handsaws also come in lesser sizes. The bucksaw handles small trees and poles. The Swedish bow saw is an improvement over the bucksaw. It has a blade that stays sharp longer and cuts faster. Don't try to use a carpenter's handsaw. It does not have teeth designed for making large cuts.

SPLITTING MAUL
AND WEDGE

Splitting tools include wedges and a maul. A sledgehammer can drive wedges, but the splitting maul has a hammer on one side and a wedge-shaped head on the other side. It is heavy enough to pound wedges into hardwood, and when swung with force, can break open a cracked log. If you are looking for hard exercise to put your body in good physical condition, splitting wood is for you. Break yourself into it gradually, though, or it will break you. The only substitute for hand splitting is a power-driven affair that pushes a wedge-shaped probe into a block of wood with hydraulic ram or screw. Such a machine is cumbersome and hard to find and probably not feasible for small operations.

A chain saw is a good investment even at two or three hundred dollars. Get some advice from people who use them before you decide on make and size. Since a good chain saw, if properly cared for, could be a lifetime investment, don't make a hasty purchase, or you will be making expensive trade-ins until you get what you want. Factors to look for are light weight, ease of starting, and length of bar. Every year, chain-saw accidents leave people mangled and maimed. It is impossible to live free from all hazards, but careless use of powered saws is a quick road to disaster. As much careful thought and planning should go into starting and operating a chain saw as does taking off with a light airplane. A chain saw is a remarkable tool that, if handled properly, can make country living a more satisfying way of life. But maybe you shouldn't neglect learning to operate the hand tools. Someday they may be all that you have

Probably the most important consideration in establishing a wilderness home, whether in a remote area or in a rural community, is water. You can improvise many things but not water. A good spring seems ideal, but even it requires proper management if it is to provide a continuous supply of clean water. The wilderness pioneers of a hundred years ago could choose their living site from a large unclaimed land, and the presence of a reliable water supply often influenced their choice. They usually built their house below a spring so they could pipe the water into the building with a gravity-flow system. If they dug a well, they situated the house as close to it as possible.

The quantity of water used was only a fraction of what most people think is necessary today. Our wilderness property has neither a spring nor a stream on it.

On our first vacation spent there I anticipated spending a large portion of my time hauling water from a nearby spring in a five-gallon container. It seemed to me that five gallons of water would last only a few hours. But I had forgotten that in wilderness living a little bit of water will go a long way. We lived in a 10-by-12-foot tent and cooked over a campfire. The tent had no sink with a faucet from which water could run lavishly. A small basin on a stump held water for cleaning hands and face. For washing and

rinsing dishes after each meal, we heated a kettle of
water. The improvised outdoor privy didn't use water
for flushing, and baths were a simple affair with a
washbasin of warm water set in a small inflatable
plastic wading pool to catch the splashes.

To my surprise, I discovered that we did not use
even five gallons a day. A fifty-gallon barrel of rainwater could keep us supplied for two weeks. Now, I
realize that year-round wilderness living involves
more than a tent and a campfire, but I would like to
emphasize that a simpler pattern of living requires
much less water than most people realize.

A permanent wilderness home need not be without running water. The greatest satisfaction comes
with solving the water problem in a way to provide
the most comfort and convenience possible but with
the least expense and a minimum of equipment and
energy. Even a small spring at a higher elevation than
the house can fill a storage tank that will in turn
supply a constant stream of clean water. The rate of
flow and the greatest quantity used in any one day
will determine the size of the tank. A tiny trickle of
water flowing all night into a tank can meet the entire
needs of the next day. A fifty-gallon barrel can hold
enough water for the daily needs of four if the spring
can fill it during the night and the portions of the day
when they draw no water. A simple plumbing arrangement to bring water from the reservoir barrel to
the kitchen sink will solve most of the water problem
for a comfortable wilderness home.

In the event that you have no spring at an elevation
above the house, a nearby well may offer the next-best
solution. Select your building site with the water table
in mind. If you can dig a well that has water in it 30
feet or less from the surface the year round, all you
need is an old-fashioned pitcher pump and a little arm
work.

The neighbor's hand pump that he had installed

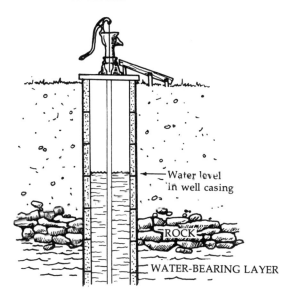

Water level
in well casing

ROCK

WATER-BEARING LAYER

right beside the kitchen sink always intrigued me as a boy. Our well stood just outside the kitchen door and required only a few steps to bring a bucket of water from the hand pump to the kitchen table, but the neighbors had built their kitchen right over the well so that the hand pump could deliver water right into the sink. Not only that, they had a special type of pump which delivered a steady stream of water instead of the intermittent splash of the pitcher pump.

The basic difference involved a small pressure reservoir built right into the pump. You pushed the handle down, drawing water up from the well which went through a small pipe that had a side chamber filled with air. Since the water could not flow rapidly out of the small opening, the volume forced some into the air chamber, compressing the air. The air pressure maintained a continuous flow of water from the spout while you raised the handle to bring up the next surge of water. I have hopes of finding one of those old pumps to install in my wilderness home.

Also I remember another ingenious hand-powered water system at the old homestead place of my friends a few miles away through the woods. Their well was a few feet from the house, and they had an elevated water tank beside the house about at roof line. Instead of the regular pitcher pump at the top of the well, theirs was a horizontal in-line pump that they worked by an upright wooden handle pulled back and forth. The pump plunger pulled water up from the well when you moved the handle toward you. Then when you pushed it the other direction, it sent the water up through a pipe and into the tank about ten feet above the ground. The tank held about sixty gallons, and water pouring out of the overflow signaled when it was full. A pipe from the lower part of the tank carried water to the kitchen sink.

When electricity arrived, of course, families modernized many of the old homesteads with an electric pump and pressure system. But when the power fails, the neighbors will often gather where the hand pump is still in place, grateful for a bucket of water to help them through the emergency.

The water problems of each homestead in the wilderness varied, and their solutions offered a tribute to the pioneers' ingenuity in providing the comforts of life with whatever they had. Have you ever seen a hydraulic ram at work? No, it's not the name of an animal that hauls water but rather a unique device that takes advantage of the energy of falling water to pump some of it to a higher elevation. If the spring or stream lies below the level of the house but has an abundant flow of water, you can employ a ram to harness some of the water to pump a supply up the hill.

My great-uncle had a hydraulic system that sent creek water to a tank on top of the hill above his house. The tank produced a constant flow to his home. The hydraulic ram transfers the energy of water flowing rapidly downhill in a pipe to a coiled steel spring. The force of the moving water tightens the spring when a valve in the pipe suddenly stops the flow of water. As the spring uncoils, it pushes a smaller volume of water uphill. The valve then lets the water in the larger pipe flow again. The water picks up full speed in a few seconds. When suddenly stopped again, its energy of momentum once more tightens the spring, and that energy pushes another small amount of water up the pipe. While an intermittent flow, if it fills a reservoir tank above the house, gravity will ensure a constant volume to the house system. The greater the flow of water, the larger the ram you can use and the more water you can supply to the house.

You can also use the wind to pump water. In the simplest form, a windmill moves the pump handle up and down to lift water from the well. Such so-called primitive systems had their problems and failures. Many times they required the skill of a mechanic to keep them operating, but that is the challenge of the wilderness. Anyone who can make money can pay the water bill, but only those who enjoy meeting a challenge with their ingenuity can bring water to the house through their own devising.

Perhaps one of the simplest and most often neglected water supply systems is that of the cistern. In its elementary form it may consist of nothing more than a rain barrel under the end of the eaves trough. The significance of the old song about sliding down the cellar door and looking down the rain barrel escapes most people today. Even as a child I do not think I fully realized the value of the old rain barrel at the corner of the woodshed as a source of clean soft water.

The water from the well was hard, and it had a

distinct mineral flavor. Unless used with care, it could form a dark scum of soap on white clothes. But the water from the rain barrel had no minerals and made nice soft suds when used to wash your hair or special white garments. We dipped from it only when we needed soft water. Probably 99 percent of the water that poured into it spilled over the sides when it was full. Think of the water that we could have stored if, instead of the rain barrel, we had had an underground cistern.

The one side of the woodshed roof that drained into the rain barrel was about 10 by 30 feet, or 300 square feet of roof surface. Our area received about 60 inches of rainfall a year. Multiply the surface area by the rainfall and divide by 1.6 to convert it to gallons, and you have over 11,000 gallons of water from that small roof in one year. According to estimates, five gallons of water a day for each individual is usually sufficient. Let's be extravagant and double that. My family of six would then have used sixty gallons a day, or 22,000 gallons for the entire year. If we had channeled the runoff from both sides of the woodshed roof into the same cistern, we would have had our year's supply of water.

Of course there would be no point in storing up a year's supply of water when it comes down from the roof quite regularly throughout the year. A cistern of 2,000-gallon capacity would have given us a one-month supply, and only in the summer months would it likely have ever dropped to less than half full. Now, remember, we are talking about household water only, not for irrigating crops.

Also, most people live in regions of less than sixty inches of rainfall a year. But with thirty inches of rainfall, you could use twice as much roof area. Localities with long dry seasons would require larger cisterns to hold enough water to last until the next rains. Most roofs collect considerable dust and dirt during a long dry spell; so you should shunt the first rainwater away from the cistern until the roof is clean.

You may need sand and charcoal filters to ensure clean water going into the cistern. A well-built cistern, usually concrete or masonry, will be watertight. Completely cover it to avoid contamination. Chlorination is a sensible safeguard, along with frequent testing for purity. Once you have stored the water, you will have to pump it out by whatever means you consider best. A hand pump should have a good foot valve that is properly sealed or covered to prevent contamination from spillage or backflow.

Up to now we have only provided the kitchen sink with cold running water. What about hot water for a bath? For many years we lived happily and healthfully with only a once-a-week warm tub bath. We poured a teakettleful of boiling water into the galvanized wash or laundry tub on the floor by the kitchen stove. Then we cooled it with enough cold water to suit our fancy. By kneeling or squatting in the tub we could soap up and rinse off. That may not sound as comfortable and convenient as a smooth porcelain bathtub, but it certainly served the purpose. However, with a little more planning and ingenuity, the wilderness home can have both hot and cold running water.

As soon as you establish a constant flow of water in a piped system, you have the prerequisite to hot water. You can fit a good-sized wood-burning kitchen cookstove with a water-heating coil in the firebox. Copper tubing bent into a coil that fits in the side of the firebox connects to a hot-water tank or reservoir. The tank, usually twenty to thirty gallons in capacity and elongated in a vertical position, sits behind or near the stove. Since warm water rises above cold water, the water will circulate from the tank through the coil.

The tank usually has no insulation; thus it helps to warm the kitchen. By placing your hand on the outside of the tank you can determine the water's temperature. If it is hot halfway down, you have a plentiful supply for doing dishes and taking a bath. I remember that when Mother was canning and had a hot fire going all day, the water got so hot in the tank that someone had to open the faucet at the sink frequently to release excess pressure. Sometimes live steam came out instead of hot water.

Living in the wilderness does not need to be uncomfortable to be satisfying. If you use your wilderness retreat only during the summer months, heating and lighting are not a great problem. However, if you develop a place in the country as a wilderness experience, you will most likely live there throughout the year—if not now, probably later. It is much easier to provide for winter insulation and heating as you build than to try to add it later after you have completed the basic structure.

A log house, if properly built, provides a considerable amount of insulation in the thickness of the wood in the walls. If you add interior paneling, it is best to include some kind of insulating material at least in the spaces between the paneling and the log. It will help to eliminate cold spots in the walls. Roofing paper overlapped in the rows of shakes or shingles also helps to prevent heat loss. A ceiling with overhead insulation in the attic will conserve a lot of heat, but many favor an open-beam type of construction because of greater usable space, such as for sleeping lofts. Sometimes you can add a rustic ceiling with insulation on the underside of the roof. Wherever you can cut down heat loss, you will save yourself time and work when it comes to obtaining an adequate fuel supply for the winter.

I would not consider building a country home of any type without a fireplace. While I realize that one can find more efficient ways to heat a house than a fireplace, still to me the fire is as much a part of wilderness living as a mountainside is to a mountain climber. And I am convinced that many people even in the suburbs and the city find a degree of wilderness experience simply by building a fire and then contemplating their accomplishment to that degree of self-sufficiency. Maybe that is why a cheery fire on a cold winter evening puts one in a contemplative mood of contentment.

The fireplace in a wilderness home must work, not just be an ornament. If you have not had any experience building fireplaces, you may want to use a metal heatilator unit to make certain it will draw properly and not smoke. Even then you must design the chimney above and around the metal unit correctly, or it still might smoke. Locate a friend who has a fireplace that works well and make sketches and measurements of all dimensions inside and out to guide you. Or, better yet, find someone who can help you construct your first fireplace properly. Should you plan to live permanently in the dwelling, it may be best in the long run to hire an experienced fireplace builder.

Most people place their fireplaces in an outside wall. While it conserves space in a small house, think of the heat lost from the bricks or stones exposed to the outside. Unless you have some good reason to heat up the whole woods, construct your fireplace as part of an inside partition and benefit from the warmth radiated from all sides of the fireplace, even after the fire goes out. If the chimney goes up through the peak of the roof, you will have more inside to give off heat.

Wood-burning heating stoves are still available and may become more common as the various heating fuels become increasingly expensive and short of supply. Some of the open-fronted stoves are really just modified fireplaces. Although attractive and pleasant to sit by, they are not as efficient as the airtight heaters with controlled draft systems. The greatest waste in wood-burning devices is the heat that goes up the chimney. Besides that, it is much easier to regulate the rate of burning in a controlled-draft heater. I remember long winter evenings at home when the whole house was comfortably warm, and we added wood to the heater only about every two hours. Once a fire burns strongly with the drafts open, closing the drafts and dampers will slow it down to an even heat and make the wood last a long time.

Some heaters have an outside jacket that circulates cooler air near the floor into the heated space between the firebox and the jacket. It also promotes a more even room temperature. If you are interested strictly in utility and not appearance, some of the longer double-chambered heaters will provide much more heat from the same amount of fuel. Some neighbors built one from two fifty-five-gallon oil drums and several pieces of stovepipe. The fire burns in the lower drum, and the flames and smoke circulate through the upper drum before going up the chimney, accomplishing somewhat the same purpose as a long horizontal stovepipe from the stove to the chimney. The wood-burning heater and the kitchen stove can both connect to the same chimney, and if the chimney extends up an inside-partition wall, it will radiate considerable warmth to the rooms on either side and up to the attic.

The kitchen stove itself can serve as a good source of heat. Properly placed, it can warm more than just the kitchen. During meal preparation, especially baking, a lot of warmth escapes from the kitchen. In the old homestead houses one of the favorite spots on a cold winter day was next to the kitchen stove. From a perch on top of the big woodbox, we children enjoyed the steamy odors of boiling potatoes and baking bread combined with the crackling warmth of the wood fire. The kitchen table was a particularly good place to spread out various projects on a winter day. It was sometimes necessary to shoo us away in order to set the table. The dog and cat also found favorite spots around and under the wood-burning heater and kitchen stove.

Proper waste disposal is as essential in the wilderness home as in the suburb or city. However, the amount of waste, especially from the kitchen, will be surprisingly small with home-grown produce. At least the waste that results you will find much easier to dispose of. Metal, glass, and plastic will be minimal. Most of the unused materials from food preparation you can recycle back to the soil either through a compost heap or by burial. Kitchen scraps can go to feed farm animals. The few cans and bottles that do accumulate one can often use in various ways around the homestead as handy containers. We never had any need for garbage pickup at our home in the country. The animals devoured digestible scraps, burnable refuse went into the fire, and nondegradable materials accumulated so slowly that I cannot remember any time that we had to haul refuse to the county dump.

Human sewage wastes do pose somewhat more of a problem because of the hazards of improper disposal. We have become so accustomed to washing everything undesirable away with water that we often fail to recognize what that does to our entire ecosystem. Just because you live on a hill with plentiful water supply does not give you complete freedom from the responsibilities of environmental pollution. Most people depend on a supply of pure water that comes to them from wilderness and rural areas. Every bit of human sewage washed away from your wilderness home becomes a contaminant that must be removed to provide water for millions. In fact, you may unwittingly pollute your own water supply if you aren't careful.

Septic tanks, if properly built and maintained, offer a good method of waste disposal. The idea is to

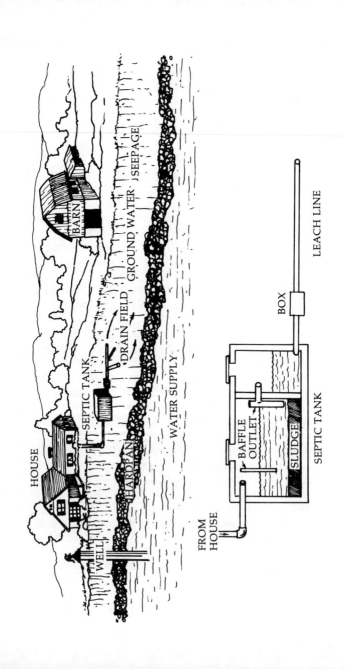

HOUSE

WELL

BARN

SEPTIC TANK

DRAIN FIELD

GROUND WATER

SEEPAGE

HARDPAN

WATER SUPPLY

FROM HOUSE

BAFFLE

OUTLET

SLUDGE

SEPTIC TANK

BOX

LEACH LINE

hold the material in a tank long enough to allow decomposing bacteria to reduce it to the same state as the decomposed leaves that make up the rich soil under the forest trees. However, too much water may wash the wastes through the septic tank before the decomposing bacteria have had a chance to break them down. The material already in solution will become part of the ground water that supplies springs and streams. It is true that if organic wastes move slowly enough through the soil, or even in a stream, the bacteria continue to work and may be able to alter them into a relatively harmless form. But don't assume that 100 feet of soil or 1,000 feet of stream will automatically purify the water. It depends on many factors such as the nature of the contaminants, the rate of water movement, the temperature, and the amount of oxygen available.

Some people extol the virtues of the old-fashioned outdoor privy for safe disposal of human wastes. A properly constructed privy or pit toilet correctly maintained will perform the same function as a septic tank, that is, to allow the complete decomposition of wastes by bacteria. But here again, a carelessly built and managed privy can pollute your own water supply and that of many others as well. In a wilderness setting or a homestead where you want to keep things as simple as possible, an outdoor privy can present a satisfactory solution to the waste-disposal problem. You must take care, however, that it functions properly throughout the year—dry season, wet season, winter, and summer. Some pit toilets consist of nothing more than cesspools or holding tanks from which contaminants continuously drain into the ground water. I remember as a child when visiting with friends, trying to delay a trip to the outhouse, hoping I could avoid a disagreeable experience. Not only were some outhouses poorly constructed, full of flies and nauseating odors, but

half full of water and without the advantages of a
septic-tank system.

The functional principle of a correctly working
privy is similar to composting. Wilderness hikers and
backpackers learn to bury human excrement shal-
lowly so that the surface warmth and soil bacteria can
promote rapid decay. A compost heap breaks down
rapidly if one mixes the organic matter with nutrients
and air, keeping it warm and moist. If a good supply
of bacteria and fungi are present, they will produce
warmth by their metabolism, especially if covered to
protect from freezing and drying or the other extreme
of excess water and no air.

With such concepts in mind, you can design a
privy that will recycle human wastes back to soil and
release a minimum of pollutants to the air and water.
If you live on a hillside, you can build a composting
pit with the privy above it. On level ground you may
have to compromise between elevating the privy and
digging a basement for it. At any rate, you must have
a well-drained pit to which you can add necessary
ingredients and remove the final product. The members of the family can soon learn that instead of pushing a handle for flushing, their responsibility is to add
a few handfuls of sawdust, shavings, shredded
leaves, or any other coarse material available. It provides a well-aerated mass in which decomposing bacteria can work efficiently. The addition of a small
amount of lime or wood ashes now and then will help
control odors, and a few shovels of soil will add bacteria and essential nutrients. Ideally, you should turn
the pile once or twice to avoid soggy spots and promote decay. Since small quantities will be easier to
handle than large ones, try a double compartment
with use alternated to the other side while you await
the completion of composting in the first side.

With proper management, a nicely constructed
building, attractively painted and well lighted, does

not have to carry the connotation of disagreeableness any more than the fancy, tiled indoor bathroom of modern houses.

Some composting systems for human wastes have been perfected to the extent that you can incorporate them into the dwelling. Your budget, your needs, your life-style, and the location of your wilderness home will determine the sanitation system best for you. Just be sure it will work properly without spoiling your environment.

What happened at your house the last time you had a power failure? In many homes it occurs so rarely that the family hasn't made any provision to deal with it. Usually a candle or a flashlight gives enough light to get by for the moment. But what was it like before the use of electricity? We have become so accustomed to flipping switches that, in most people's minds today, living without electricity is something akin to the stone age. Really, it isn't all that bad. Talk to someone raised in a home with no electricity, and you will realize that life can go on quite normally without it.

First, let's consider lighting. The simplest and perhaps most reliable light is the kerosene lamp. The flame is adjustable from low to high and is enclosed in

KEROSENE LAMP

a glass chimney to minimize flickering. It holds enough kerosene to burn for many hours with no attention. You can carry it lighted from room to room with relatively little danger, unless you stub your toe and drop the lamp. Even then, kerosene will likely flood the burning wick, and it will go out. Kerosene is not as highly flammable as gasoline. Of course, the kerosene lamp gives a rather weak light, but I have spent many hours reading by kerosene lamplight. The lamp must be within a few feet of your book, and it is usually desirable to have several lamps in the house to distribute light where needed.

Kerosene lamps are readily available today. Unfortunately the most popular are not the most economical, either for amount of light produced or for the cost of the fixture. Apparently people buy them for ornaments, and the more decorative, the better, it seems. Fancy bowls and shades with ornate glass and colored flowers do not produce much light, but they do add a dimension in old-fashioned decor. If that is all you desire, then pick out the fanciest. Grandmother did the same thing when she wanted to add a touch of elegant beauty to her parlor. What could make a fancy vase more beautiful than a softly glowing light within?

Modern kerosene lamps also come with colored kerosene to match the upholstery. The kerosene is even scented with anything from attar of roses to bayberry. If you are willing to pay several dollars per quart for such fancy fuel, you are no doubt more interested in decoration than utility. Plain unadulterated kerosene sells at the service station by the gallon at a relatively moderate cost. The reasonable prices of simple utility-style kerosene lamps have pleasantly surprised me.

They are simple pieces of equipment and, except for the fact that they contain breakable glass, not much can go wrong with them. The base holds the kerosene. The middle section is a light-metal wick holder which supports the glass chimney. The wick, made of braided cotton string, carries the kerosene to its burning edge by capillary action. A simple knob raises and lowers the wick to adjust the flame size. Expose too much wick, and the flame will burn high with much black soot. Turned too low, it will not reach above the edges of the flame guard and will, therefore, produce little light. The only maintenance necessary is refilling the reservoir with kerosene, trimming the wick for an even flame, and cleaning the glass chimney to get the most light.

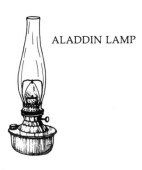

ALADDIN LAMP

Kerosene lamps come in various sizes. The larger ones have a wider wick for a larger flame. If the simple kerosene lamps do not give you enough light, then you may want to step up to a mantle lamp. One old standby still available is the Aladdin lamp. Instead of just a simple wick, it has a cone-shaped mantle that glows in the flame. The principle is the same as the gas-fired camping lanterns sold by Coleman and others, except, in the Aladdin, the fuel is just kerosene and it has no pressurized system; also, it is much simpler to operate and maintain, and, if properly adjusted, it will produce the same illumination as a 75-watt electric light bulb. One such lamp in the middle of the table or hung from the ceiling provides a warm glowing light that spreads to the edges of a good-sized room.

A number of other light sources do not require electricity. The Coleman lamp is a dressed-up version of the camp lantern. The mantle, large and cone shaped, gives off an intense white light from the heat of the gas flame. While most such lamps require highly refined white gas or special lantern fuel, some will burn regular grade gasoline. The fuel must be carburated under pressure to produce a mixture of gas and air that will burn clean and hot—which requires clean fuel and properly adjusted jet orifices.

If you are mechanically oriented and enjoy tinkering with gasoline engines, then you will probably feel right at home with pressurized gas lamps and lanterns. You will have to oil the pumps, clean the generators, and replace pressure caps. Getting one to light may often resemble troubleshooting a balky gasoline engine. On the other hand, almost anyone can learn to service and light a kerosene lamp. Dirty water and sand may pollute the kerosene, but the lamp will still give off light at the strike of a match.

No country place should ever be without a kerosene lantern—the outdoor counterpart of the kerosene lamp. Although anything but glamorous, it

KEROSENE LANTERN

is one of the most useful light sources I have ever used. Metal replaces glass for everything but the chimney, and even then metal posts and top guard it. The draft holes and chimney top protect it against gusty wind. Even rain and snow do not stop the steady light from the old barn lantern. Carry it by a wire bail or handle and swing it at the side like a bucket while walking. When you need both hands, set it on the ground or workbench without fear of rolling or tipping it over. A pleasant companion on a winter evening at chore time, it gives off enough heat to warm cold hands held over the top and a warm golden light that radiates in all directions, not in just one like a flashlight. You can temporarily hang it from a nail or a wire to illuminate a room, and it will burn for eight to ten hours on one filling of kerosene.

A few evenings ago, I went on an errand with my kerosene lantern in my hand and happened by the neighboring farmyard. There I noticed someone trying to hitch the big tractor to a piece of equipment in the dark. I sensed frustration as a fumbled flashlight kept pointing at everything but the hitch, and the man needed both hands to make the proper connection. I stepped over to the scene and asked, as unobtrusively as possible, "Would you like some light?"

"Yes, I sure would!" he immediately answered. It was easy to see also that four hands would do the job much more readily than two, so I set the lantern on the ground and proceeded to help. We completed the hookup in a few seconds.

"Say, that lantern is all right!" my friend observed.

"Yes, it is handy; you ought to have one around," I replied.

"I think I do have an old one somewhere out in the shop," he mused.

While I bit my tongue to keep from saying something like, "You dumbhead, why don't you drag it out?" he fumbled with his flashlight and exclaimed,

"You know, I think I'll see if I can find it. Thanks for the idea."

I wonder how many other old barn lanterns gather dust when someone could put them into good service. Some people object to the fumes from kerosene lamps and lanterns and challenge them on the basis of air pollution. I would encourage any such person to consider the pollution resulting from the generation of electricity or the manufacture of disposable flashlight batteries.

In the past ten years I think I have thrown away at least a dozen flashlights that refuse to come on or stay on after a short period of use, not to mention the number of batteries that had worn out. True, burning kerosene produces a pungent odor, especially in a tightly closed room. It would take some readjusting to become accustomed to it again, but before electricity, people didn't worry about that any more than they did about the odor of burning grease from the frying pan.

We talk about using solar energy. Kerosene is solar energy trapped by green plants a long time ago. It is a nonrenewable resource, but we don't need to hoard it—just use it with good sense.

Could you get along without a Deepfreeze or even a refrigerator? Not having a Deepfreeze simply means more canning, drying, and careful planning. But no refrigerator—that might pose a somewhat greater problem. I have difficulty imagining what we would do now without one until I remember that we had none in our house until I was fifteen years old. We didn't even have an icebox because of the difficulty of obtaining ice. What we did have was a cupboard in the north wall of the kitchen which had a screened opening to the outside. During most of the year the outside air temperature was cooler than inside, so it was the best place to put food we wanted to keep over to the next meal.

In the warm summertime we had an abundance of fresh food from the garden, and leftovers usually went to the chickens and other animals. In the wintertime when it was more essential to save leftovers, the cooler provided sufficiently low temperatures to preserve most things. Since the cow provided fresh milk and cream twice a day, any left over we allowed to sour for cottage cheese and butter. Ice cream we made in a hand-cranked freezer when ice was available. If some remained uneaten, we placed the container back in the crushed ice remaining in the freezer and covered up the whole thing with an old rug or blanket. It would easily stay frozen overnight and maybe longer, depending on the air temperature.

Of course TV dinners and frozen foods would not do in a household with no electricity. Yet people seemed to get along without them once, and with proper adjustments, I'm sure they would again. A lack of refrigeration complicates the preparation of proper diets with balanced nutrients, but with all of the diet information available today, it seems likely that one could solve the problem more easily now than fifty years ago. Dried and canned foods may lack some nutrients found in fresh frozen foods. But knowing which nutrients are missing and where one can obtain them helps in planning a balanced diet without the aid of refrigerators and deep freezers.

Many homes already use some source of energy other than electricity for cooking and heating. Although gas and oil are almost as convenient as electricity, I suppose the ultimate in independence is a wood-burning stove, but I wonder how many people who talk about getting one could make the stove work if they had it. It's not just a simple thing like building a fire in the fireplace. Merely starting a fire requires the proper use of kindling and fuel wood, and to keep it going involves the correct adjustment of drafts and dampers.

It would be interesting to have a contest to see who could get a teakettleful of water boiling the quickest, or who could keep it boiling steadily for ten minutes. Even more challenging would be competition to bake a cake in the oven. Watching someone who has had many years of experience with a wood-burning stove may lead you to think it is a simple thing to operate. You will no doubt change your mind when you try it yourself the first time. I don't mean to discourage anyone from acquiring a wood stove—the experience can be pleasant and satisfying if he is ready to learn a new skill.

Your wood supply can make the difference between fun and frustration. Damp wood, or wood that does not burn easily, can bring meal preparation to a standstill. Housewives had many words to say to negligent husbands who failed to provide them with adequate fuel for cooking. It takes some planning ahead. Usually one cuts cookstove wood green, then he must cure and store it properly. Vary the size. Small pieces speed up a slow fire, and large pieces hold it steady. Most housewives had to develop some ability to split wood into the proper form for the immediate needs because a batch of bread only half-done couldn't wait for someone to come home from work to chop more. Some kinds of wood are easy to split, but others demand a real skill in wielding an ax and maybe splitting wedges also.

Maintaining a wood-burning kitchen stove involves more than just putting in wood and working the dampers and drafts. You must thoroughly shake the ashes through the grate into the ash container underneath the firebox so the draft air can reach the fire.

When the ash box fills, you must empty it. The best time to do that is when the stove is cold, or before you kindle a new fire. A box of hot ashes is pretty tricky to handle. What's more, you don't usually want to stop in the middle of meal preparation to empty the ashes. Cold ashes you can spread on unused garden ground, but hot ashes with live embers might start another fire where you don't want it. If the wood had a considerable amount of pitch, you will have to clean the black soot out from the smoke chamber that surrounds the oven box. If too much soot accumulates, it will act as an insulating blanket and prevent the oven from heating properly.

The tools that go with a kitchen stove include, along with the lid lifter and grate crank, a long-handled soot scraper that you insert through a small door below the oven to pull out soot and ashes pushed down from the top. A fire that gets too hot, especially if the extreme heat concentrates in one spot, can seriously damage the stove. The top can warp and even crack, and the grate can burn out. In a damp climate the bare metal parts tend to rust. Stove polish will keep them clean and shiny. A well-cared-for wood-burning stove can be an attractive part of the kitchen equipment that will last for many years.

CHAPTER 9

FARM ANIMALS

Producing your own food in the country may or may not involve farm animals. Nutritional research has established the fact that you can obtain a balanced diet from plant sources, but as long as milk and eggs are relatively free from disease, many people will continue to use them.

Some people are so fond of horses, cows, and chickens that they would have them on their country place whether or not they used horses for work and milk and eggs for food. If you have not decided whether to raise animals yet, you should certainly consider the life-style that goes with animals that require constant care. I remember my childhood unhappiness when we wanted to go somewhere overnight and couldn't because someone had to milk the cow night and morning without exception. I was as grateful as the rest of the family for milk, cream, butter, and cheese to eat, but I found it hard at times to reconcile myself to the fact that we had to follow the cow's schedule instead of our own. Unless you really like that kind of life, you had better consider the possibility of buying your dairy products from someone who makes a living at dairying.

Consider a simple cost analysis. Keep count of how much you spend for dairy products for one week and multiply by fifty-two to determine your total yearly expenditure. Then figure the cost of maintaining a cow for a year and see how the two compare. Besides the expense for dairy feed and hay for the year, you should add a reasonable fraction of the initial cost of the cow and barn and milk-handling equipment. Even should you have enough land to produce your own hay and pasture, it will take some equipment to cut and harvest the hay. You may not want to consider the cost of your time, but you should include something for the value of the property, the taxes, and the fencing. In addition, you may have veterinary bills and the cost of medication for breeding and calving.

On the other side of the ledger you should consider the value of having as much milk, cream, butter, and cheese as you want and of knowing that someone has properly cared for and processed it. It is difficult to place a dollar value on the satisfaction of producing your own dairy products and on the pleasure derived from taking good care of a family cow. The cow will often become one of the family pets and provide excellent training for the children in how to love and care for animals.

The various breeds of milch cows have their advantages and disadvantages as family pets. Holsteins give the largest quantity of milk and one of the lowest in fat, but they are large animals and not easy to manage. Guernseys give a moderate amount of milk, higher in fat, and they may or may not be easy to manage. Usually easier to manage, being smaller in size and more adaptable in disposition, are the Jerseys, which give the smallest amount of milk, high in fat. The various other breeds fit somewhere among the three mentioned.

In spite of my childish frustrations with being tied to the cow's schedule, I can recall with considerable pleasure the walk to the pasture in the evening to bring the cow in for milking, the pleasant aromas of fresh hay and warm milk, and the musical sound of milk streaming into the bucket between my knees. Although we never kept more than one cow, due to our limited space and feed, we always had an abundance of milk, cream, butter, and cheese. During the season of peak production we even had a small surplus to sell or trade. With only one cow, we went without dairy products during the six to eight weeks just prior to calving each year when we had to dry up the cow and feed it extra well to produce a healthy calf. During the dry period we usually purchased some milk from the neighbors.

The barn was small but well built and adequate for the needs of one cow. Most of it served for hay storage. The stall was just wide enough for the cow and the milking stool. A single stanchion and manger held the cow during milking, and a small pen beside the milking stall helped to separate the calf from the cow during milking time. We kept the calf only long enough to take advantage of the excess flow of milk after its birth. Most of the times we sold it to someone who had several cows and wanted a calf to fatten for beef.

Usually we had enough hay stored in the barn to provide as much as the cow could eat during the winter when the pasture did not produce much food. A few pounds of dairy feed each day with a few mangel beets and surplus vegetables from the garden supplemented the hay. Since a cow requires large quantities of water, especially while eating mostly dry food and producing up to three gallons of milk each day, we kept a large tub in the barnyard full.

The barn we cleaned at least once each day and threw the manure in a stack to rot prior to placing it on the garden. It contained a considerable amount of straw and uneaten hay from the bedding in the stall. The mixture produced a compostlike material that improved our poor garden soil tremendously. We usually left the manure in the stack through the winter rains and spread it on the ground in the spring. That way it retained the maximum nutrients. The stack could shed rainfall enough to prevent severe leaching. In considering the benefits derived from keeping your own cow, don't overlook the excellent fertilizer and soil conditioners it produces year round.

A cow will learn to lead quite easily with a rope and halter. You must train her early to respond to commands, however, and she may require some disciplining. As long as she recognizes authority in humans, she will respond to commands. But if she once learns she can have her own way, she may become almost impossible to handle

Chickens are good protein-producers if you use their eggs. Man has artificially selected the domestic chicken to produce large numbers of eggs during a longer part of the year. Many different breeds of chickens exist, some better suited to the small farm than others. Breeders produced the white Leghorn mainly for large commercial flocks of egg layers. But Rhode Island Reds, New Hampshires, and Plymouth Rocks are just three of the many varieties well suited to backyard or barnyard living. Chickens are easier to keep than cows, but they too demand almost complete attention, with no breaks in schedule. They must have food, water, and protection from extreme weather to remain healthy. Besides the standard egg-laying breeds, you find novelties like the Bantams and silkies which may produce a few eggs, but people keep them mainly as curiosities.

Unless you want to spend a lot of time looking for hidden nests full of eggs, you should provide a permanent house with nesting boxes for the chickens. Make the nesting boxes dark inside with some straw or other cushioning material to prevent excess breakage of eggs. If you build them on the side of the chicken house, you may be able to gather the eggs from the outside, thus avoiding dust and dirty shoes. You do not have to include roosters with the hens to raise egg production. Many chickens will lay 250 eggs in one year, but the length of daylight, which constantly changes, influences the egg laying. The wild chicken or bird lays all of its eggs in the spring of the year, when the daylight hours are the longest. Modern animal scientists have capitalized on the pattern by employing artificial lights in the winter to fool the hens into thinking it is spring and egg-laying time. You can follow the practice to a certain extent on even a small flock. However, you will need to get up early to turn the lights on, and don't forget to flip them off before you go to bed at night. Better yet, if you can find an automatic timer, set it and forget it.

Most homestead farms will want only enough chickens to provide eggs for the family. Many nutritionists now tell us that we should eat no more than one or two eggs a week. At that rate, one good laying hen will satisfy a family of four. But who wants to keep just one lonesome hen around? Maybe you should figure up the total expense of a few chickens and decide on the basis of whether or not you can buy the eggs you eat for less than you can raise them yourself. However, eggs are easy to sell, and a small flock of chickens might give you spending money for some things you wouldn't have otherwise. To avoid expensive feed, let the chickens run outside to gather as much of their own food as possible. Otherwise your expenditure will be greater than your income. Five-dollars-a-week income from eggs sounds pretty good until you realize that you spend four dollars a week for feed in addition to your investment of time and equipment. But if you can grow most of your feed, then the project becomes much more financially sound.

Day-old chicks are the most economical to start with. They can travel through the mail because they don't eat for a day or two after hatching. During that time they still subsist on the supply of food stored up in the yolk of the egg from which they hatched. The little animated fuzz balls are a delightful addition to any household, and their loud cheeping will get them plenty of attention—which they require for the first few days, without a mother to care for them. In the place of the natural mother, you must supply a warm place near a heater or in a specially built brooder for them. They will need water right away, and it is surprising how soon they start to scratch and peck at the fine chick feed you give them. For two weeks they will need 90-degree temperature and 80 degrees for two more. Let them out-of-doors on warm days after the first two weeks. Have a brooder going for them for seven weeks at nights and on cold days. Most feedstores can sell you chicken scratch and starting mash.

Your chicken house should be dry, light, and well ventilated but free from drafts. Plan on about 4 square feet of floor space for each bird. You will need feed troughs that allow the chickens to eat mash but not scratch it out onto the floor. Scratch feed, usually a mix of cracked corn and wheat, you scatter on the straw-covered floor where the chickens will find it. Have clean water available at all times, and a roosting place above the floor is mandatory. Nest boxes above the floor will assure cleaner eggs with fewer cracks.

The chicken litter can safely accumulate on the floor up to a depth of 4 inches. If kept dry, you do not need to clean it out more than twice a year. The droppings under the roost, however, require daily removal unless they fall into a screened pit where they can collect for as long as one week. The droppings and the litter, composted together, offer an excellent source of fertilizer. Each chicken will produce over one hundred pounds of manure during a year's time.

Start your chicken venture cautiously with ten to twenty hens until you convince yourself that it is something that you enjoy and can succeed at. Then you may increase your flock to make the time spent more productive. If you plan to sell eggs, it doesn't take much more time to care for fifty hens than for ten.

Any list of farm animals should include honeybees. Not everyone likes to work with them, but if you are interested in learning, you will find that it can be a rewarding as well as fascinating activity. In addition to honey for your table, the bees pollinate your orchard and garden. Unless you have neighbors with bees not more than half a mile away, you should really have some of your own for pollination, if for no other reason. Some beekeepers will rent their bees to you during the blossom season to pollinate your crop. However, it means a cash outflow and a complete loss to you of the nectar flow from your trees which goes to make honey for someone else.

Bees, like other farm animals, require some expenditure to get started and also to maintain. However, they are much simpler to manage because they gather all their own food from a source that you couldn't use in any other way. Like any other farm-animal project, you should start small and then increase as you learn.

Biologists as well as beekeepers have written books about the insect. Scientists have studied their habits and social life in great detail. Read as much as you can both about their habits and how to manage them.

Though the initial investment in bees will be quite high, the equipment will last for many years, and the proceeds will pay it off rapidly. The beginner can purchase a complete kit to get started with one colony of bees. It includes housing for the bees, tools and equipment for the beekeeper, and the bees themselves.

Honeybees are not native to North America, but those that have escaped domestication will select a hollow tree with a small opening in which to hide their supplies of food and growing colony. Have you ever tried to rob the honey from a wild swarm of bees in a bee tree? It's done, but usually it involves the death of all of the bees in the colony and a lot of work and misery for the people involved. Still, the romance of finding a bee tree and reaping a bountiful supply of free honey led us boys into several such adventures.

We found out how to get the bees to tell us the whereabouts of their hiding place. First we placed a small dish of honey-soaked bread near a watering place where the bees came to drink. The bees soon found it, and we watched them carefully to determine the direction in which they flew away. Then we followed with our dish for 1,000 feet or so and sat down beside some flowers to wait for more bees to find our offering. If after feeding they took off in the same direction, we knew we were on the right track. But if they flew in the opposite direction, we assumed we had passed their hiding place. After several moves we would consider ourselves in the general vicinity and begin a careful search. More often than not the bees had led us to a domestic colony, maybe on our own place or the neighbors'. But once in a while we would actually find the secret hiding place of a wild swarm. Our scheming and planning for a raid went into full gear.

Usually the opening was too small and too high above the ground, the tree was too large and too hard to cut down, and the bees too vicious, so we contented ourselves with watching them and guessing as to how much and what flavor of honey was inside. On one occasion, though, we decided to attack. The tree, as I remember, was a dead, pitchy fir about 2 feet in diameter. We borrowed a falling saw and even thought of such details as a bottle of saw oil to keep it running smoothly in the pitchy wood. Although the bee entrance was several feet above our heads, our activities soon brought the wary guards buzzing around us.

Knowing that a little smoke would help, we kindled a smudgy fire at the base of the tree. The pitchy sawdust fed the fire faster than we had anticipated, but we kept the long hand-pulled saw going with liberal amounts of oil, because it seemed that a combination of charcoal and pitch on the warm saw blade was making it particularly sticky. After a frantic endeavor to keep the fire under control, the saw moving, and the beestings at a minimum, we did manage to fell the tree. When the air cleared of smoke and bees enough for us to see, we discovered, much to our dismay, that we had cut right through the center of the honeycombs. That sticky saw was not so much due to pitch as honey oozing from the cut. Still we gathered up a considerable quantity of crushed, dripping combs in a large washtub, along with a generous portion of scorched moss, bark, and saw oil, plus a good dose of beestings. Which one of us would ever admit to anything but that the honey was of the finest flavor and quality ever gathered?

The modern beehive makes robbing the colony safe for the beekeeper and harmless to the bees. There is no need for a single beesting or dead bee if one does the operation properly and carefully. The bottom board forms a base for the stacked hive boxes and provides an entrance-exit opening for the bees. Each box contains ten light wooden frames which enclose the honeycomb so that you can lift it out one frameful at a time. The lowest box (or two boxes) is the hive body or brood chamber. In its combs the queen lays thousands of eggs. They hatch into hungry larvae that are fed on a mixture of honey and pollen. In the upper boxes, called supers, bees fill the frames with pure honey. It is those frames that one takes to the extractor to have the honey removed.

As long as she has empty combs, the queen keeps on laying eggs in the brood chamber. The colony may grow to fifty thousand, even a hundred thousand, bees. Each female worker bee makes hundreds of trips to the blossoms to gather nectar and pollen. She stores the nectar in the cells of the comb. After it is properly cured, a bee caps the cells with wax. The bees will not stop until either the nectar supply fails or they have filled every cell. A strong swarm will store away two hundred to three hundred pounds of honey in a good season, and fifty to one hundred pounds of that is surplus beyond its winter needs. The beekeeper can take the surplus, leaving the rest for the bees, or he can remove an additional portion of the bees' supply if he will feed them enough sugar water. When honey sells for a higher price than the cost of the sugar water, it is good business. On a small scale, the farmer who keeps a few colonies of bees for his own use will do best to take only the surplus and not have to worry about feeding his bees. Even at fifty pounds a colony, five colonies would provide plenty of honey for the average family. My four colonies work in an area considered poor in honey production, and I rob them lightly, but we have plenty for our own use and some to give to friends.

The beginner's kit will contain, in addition to the housing for the bees, a head veil and gloves for the beekeeper to wear and a smoker and hive tools to use in opening the supers. With the addition of a two- to three-pound package of bees including a young queen which he can order by mail, the novice is ready to begin.

BEEKEEPING EQUIPMENT

Get your hive completely assembled before the bees arrive in the mail about blossom time in the spring. Read as much as you can about how to handle the equipment and the bees and then launch out to gain experience. Like every other hobby it is only by experience that you really learn the art of beekeeping. If you have an experienced beekeeper friend or neighbor, offer your services as an apprentice, and you will save yourself some frustrating mistakes later on. Once you get your own colony working for you, you will need a capping knife and an extractor to harvest your share of the honey. Several beekeepers can share the tools to save on the initial investment.

Bees can bring you some cash income as readily as any other farm animals. As you enlarge your number of colonies, you will have more honey to sell and more bees to rent out as pollinators. But you will have to stop expanding at some point unless you want to become a commercial beekeeper to the exclusion of all of the other aspects of country living.

Even if you venture on a small scale, you should be aware of some of the hazards and problems you face in maintaining the health of a colony. Disease is always the specter that haunts domestic plants and animals. In their natural state, diseases kill off all the weaker and less-vigorous swarms of honeybees, leaving only the strongest and healthiest to propagate their kind. Disease might destroy 90 percent of the wild colonies in a large area, but after a time healthy colonies will repopulate the area. You will be managing your bees in an artificial environment, several colonies crowded together and protected from natural enemies. Since you cannot tolerate a sudden loss from disease, you must do everything possible to protect them.

Every county in beekeeping areas has a bee inspector paid to examine your bees at least once a year to make certain they are free from disease. For that reason you must register your hives with the county agricultural commission and keep them informed as to the number of colonies and their location. Feel free to call the commissioner's office for information about diseases and their prevalence in your area. Foul brood is the one that you must watch out for most. Maintaining clean, healthy colonies and properly disposing of any that do become diseased will minimize it. It is the people who fail to cooperate in such a disease-control program that make it difficult for others. One careless beekeeper can infect the whole countryside with spores of the deadly foul-brood organism.

You can detect foul brood by a rotten odor in the combs, but by then it may be too late to save the colony. The bee inspector might have spotted it in time to help you treat the colony properly to save it. Large commercial beekeepers frequently have to use preventive medications. You can avoid disease in a few colonies by sanitary measures and wise management. The bees will have an easier time keeping their house clean if you keep it in good repair for them. If you don't rob them too closely or give them too many empty combs to take care of, they stay healthier.

Protect the hive from excessive heat and cold and have proper air circulation. If you are going to expose them to full sunlight, paint the boxes white to reflect the heat or put a shade over them during the hottest weather. The bees have an efficient air-conditioning system, but it can function well only in a correctly built hive. When the temperature increases to where the beeswax might soften and sag, allowing the honey to spill, the insects take up positions at the hive entrance. By fanning their wings, they circulate air through the hive to cool it.

They will also haul water to the hive and spread it over the surface to evaporate, thus cooling the combs much like an evaporative cooler works in your home. Make certain you have a good supply of clean water available to your bees during hot weather and near enough so they don't have to expend all of their time and energy hauling it. I can always tell when my hives are getting overheated by the extra-large number of bees at the birdbath. They haul water in the first honey stomach and empty it out when they reach the hive. Think of the number of trips it must take to cool the hive on a day when the thermometer reads 105 degrees in the shade.

During the cold winter months the bees warm the hive by burning the fuel they stored up during the previous season. Just as we maintain the heat in our bodies by metabolizing the carbohydate food we eat, the bees eat honey and produce heat from it. By clustering together in the center of the hive, they can survive even though the thermometer outside the hive registers 20 degrees below freezing. One of the by-products of carbohydrate metabolism is water. As the bees maintain their body temperature, they give off water vapor that condenses on the colder parts of the comb and hive unless the air circulates properly. It is important, therefore, to allow for air circulation in the hive even in the coldest weather. Sometimes people cover or wrap the hives with tar paper to insulate them from the cold. When doing this, leave a small opening near the top.

After the long winter has passed and warmer weather brings renewed activity in the hive, you must provide plenty of space for the development of a new brood. If you removed the honey supers for the winter, leaving only the brood chambers and the winter honey supply, the hive will have crowded conditions when the queen starts laying eggs for the new season of workers. If it persists for too long, the workers will build some extra-large queen cells and begin feeding the larvae in them a steady diet of royal jelly produced by the nurse bees. The special larvae grow large and will later emerge as potential queens for the colony. As soon as they are well on the way, about 60 percent of the workers take off in a swarm with the old queen to establish a new colony somewhere else. Should you witness it, it will startle you at the suddenness of it all.

Some signal sets off the exodus, and the bees pour out of the hive to follow the queen, who flies to a nearby shrub or tree and rests while scouts go in search of a new home. The thousands of workers, each with a stomachful of honey, surround the queen in a living mass to protect her. As soon as it locates a new home, the swarm will move on. But before they do, you have a chance to provide them with a new home and invite them to stay and work for you. Have an empty hive ready for such a moment. If the bees are where you can get to them, gently shake them into a box with a few empty combs in it and hope they will decide to make that their new home. You can make it seem more inviting by giving them a frame or two of brood from another swarm. Caring for the larvae will give them work to do at once.

I found a swarm from one of my colonies on a branch of a nearby bush. The clustered bees bent it to the ground. The simplest thing, I reasoned, would be to raise the branch up enough to slip the empty box under the swarm and shake them into it. When I tried to lift it, however, the entire swarm dropped off onto the ground. I placed the cover on the box and moved it close to the mass of bees on the ground. Then, as if by magic command, they began to march through the front door into their new home. I have never seen a more beautiful sight than that golden stream of life flowing smoothly into the hive. Somewhere in the ranks marched her royal highness, the queen, even though I could not see her. Once all the bees went inside, I closed the opening. After dark I moved the new hive to its place beside the others, where it developed into a flourishing new colony.

In the old colony we find a crisis of a different sort.

From the several well-developed queen cells only one queen must come to rule the colony. The worker bees wait patiently until a queen emerges to rule their kingdom. She will search out all of the other queen cells and sting the occupants to death. If one of them should appear at the same time or before she has a chance to do away with it, the two will battle until one dies.

Other farm animals are also worth their hire. Some people prefer goats to cows for dairy products. If you enjoy their personality and the flavor of their milk, they will serve you well. Goats are somewhat more efficient, and they can thrive on a more varied diet. However, you will find that the quantity and quality of the nanny's milk is directly related to the food you provide for her.

Horses have become more important as recreational animals and pets in recent years than for working a farm. Fuel shortages may start a trend back toward farming with horsepower. Here, as with all other farm animals, you should carefully evaluate the cost of purchasing and maintaining workhorses and equipment as compared to powered machinery for the same job. As long as fuel is available, tractors will probably remain more economical. A tractor burns fuel only while in operation, while a horse must eat all of the time—and food for the horse is no longer cheap. Horses are still an attractive addition to a larger farm, however, and one can justify them on the basis of pleasure wherever the budget will allow it.

CHAPTER 10

FILLING THE SHELVES

Besides learning how to produce your own food, you will also need to know how to store it away safely for the winter months. Supermarkets and refrigerated transportation have all but eliminated the need for canning, freezing, and drying at home. But what will you do if you suddenly do not have access to the food in the supermarket?

Forty or fifty years ago supermarkets did not exist, and people provided for themselves to a much greater degree than they do now. My mother had 650 canning jars, and she always filled them by the end of the summer. Some jars she used a second time during the year when fruits and vegetables were available. About 800 quarts of fruits and vegetables represented a good supply for a family of six.

The basic principles of food preservation are somewhat the reverse of composting. The latter promotes the breakdown of organic matter by bacteria through warmth, food, air, and water. To keep bacteria from spoiling food, we must deprive them of their basic needs or get rid of the bacteria. Most canning processes elminate bacteria by sterilization with heat and protection of the food from reinfection. Freezing deprives bacteria of the essential warmth for growth, and drying removes the water essential for their lives.

One may can by either the cold- or hot-pack method. The hot-pack involves cooking the food and placing it in sterilized, sealed jars to protect it from

STORAGE OF VEGETABLES AND FRUITS

Commodity	Place to Store	Humidity	Length of Storage Period
Vegetables:			
Dry beans and peas	Any cool, dry place	Dry	As long as desired
Late cabbage	Pit, trench, or outdoor cellar	Moderately moist	Through late fall and winter
Cauliflower	Storage cellar	Same as above	6 to 8 weeks
Late celery	Pit or trench; roots in soil in storage cellar	Same as above	Through late fall and winter
Endive	Roots in soil in storage cellar	Same as above	2 to 3 months
Onions	Any cool, dry place	Dry	Through fall and winter
Parsnips	Where they grew, or in storage cellar	Moist	Same as above
Peppers	Unheated basement or room	Moderately moist	2 to 3 weeks
Potatoes	Pit or in storage cellar	Same as above	Through fall and winter
Pumpkins and squashes	Home cellar or basement	Moderately dry	Same as above
Root crops (miscellaneous)	Pit or in storage cellar	Moist	Same as above
Sweet potatoes	Home cellar or basement	Moderately dry	Same as above
Tomatoes (mature, green)	Same as above	Same as above	4 to 6 weeks
Fruits:			
Apples	Fruit storage cellar	Moderately moist	Through fall and winter
Grapefruit	Same as above	Same as above	4 to 6 weeks
Grapes	Same as above	Same as above	1 to 2 months
Oranges	Same as above	Same as above	4 to 6 weeks
Pears	Same as above	Same as above	4 to 6 weeks

spoilage bacteria. The cold-pack method puts the food in jars before cooking and then boils the full jars in a water bath to destroy bacteria and to seal the lids. Small fruits like berries are easier to "hot pack" because you can pour them into the jars while boiling hot. Large fruits like peaches and pears are easier to handle with cold packing.

Most fruits are relatively high in acid and do not provide the proper medium for many of the dangerous spoilage organisms like botulism. Vegetables and meats, however, are less acid in reaction and thus more likely to support the growth of bacteria and other decay organisms. For that reason you must process vegetables and meats at much higher temperatures, such as those attained in pressure cookers. Some hazard exists even with foods canned by commercial processes in metal containers. If you find a can of peas or beans with the ends of the container bulged outward, you can be quite certain that it has spoiled, and you should not try to eat it. Decay organisms produce gases that build up a pressure in the can, forcing the ends out. Of course you may notice other symptoms, including unusual color, odor, and flavor.

Home-canning jars usually have metal lids that sink inward after processing and sealing. It helps you to know that the jar of food has been properly prepared. Later it will also warn you of spoilage by bulging upward instead of downward. It would probably be best to start out with fruits, leaving vegetables for a time until you have gained some experience. The various books and pamphlets available will aid you in avoiding some of the common pitfalls. When you purchase canning jars and lids, look for a booklet by the companies that manufacture them. It will contain most of the information you will need for successful home canning. The rest comes mainly by experience.

Be sure your jars are free of nicks and blemishes on the sealing edge, then use new lids that have a smooth ring of soft sealer that will make good contact with the glass jar. Put as much fruit in each jar as possible to conserve space, but leave about three-fourths-inch space at the top to allow for expansion when heating. Make sure the edge of the jar is clean before placing the lid in position and don't process so hard or so long that some of the fruit boils over and spoils the seal between lid and jar. A high concentration of sugar in the canned fruit will discourage bacteria. Jellies and jams, for example, will keep longer after opening and exposing them to air and warmth than will fruit canned with little sugar. However, fruit that you eat in any quantity would not be as healthful with so much sugar. You may can fruit without added sugar, but it will tend to soften and spoil more quickly. The type of sugar doesn't make any difference. Cane sugar, beet sugar, or honey will all work.

Freezing has some definite advantages over canning. The appearance and nutrition of frozen foods stay closer to that of fresh than when canned. Color, texture, and flavor are more nearly natural, and you have less loss of vitamins. Low-acid foods such as vegetables and meats present less of a hazard from botulism when you freeze them instead of canning them. Freezing prevents spoilage by depriving microorganisms of the warmth essential for their growth. Therefore, the quicker you freeze the food, the more perfectly you will preserve it.

Drying can substitute for freezing. Instead of depriving the microorganisms of the essential warmth, it removes water to the extent that they can no longer survive. Many foods lose certain nutrients to a greater extent when dried than when frozen. However, dried foods are still nutritious and can be an important part of the winter supply. Apples are perhaps one of the easiest fruits to dry. You can eat them dry or replace the water by soaking and cooking to make applesauce. Since apples contain large quantities of water when fresh, to rehydrate requires equally large amounts. The dried apples are concentrated and take up less room. In fact, you may even want to eat more of them, so be careful.

I remember vividly how my brother and I used to like to fill our pockets with dried apple slices as we started on a walk to the river. We probably had the equivalent of four or five apples in one pocketful. By the time we reached the river, we had usually eaten them all. Being thirsty, we would enjoy a big drink of water. Then followed an agonizing period of expansion, and we thought we would pop. But next time we would do the same thing because the apples tasted so good.

You can dry any variety of apple, but the degree of ripeness influences the flavor. The apples should be fully ripe but still crisp, not mealy. We quarter, core, and peel them, then slice each quarter into four lengthwise slices. If you have warm sunny days after the apples ripen, you can dry them in the sun. Build a drying rack for good air circulation and cover them with cheesecloth to keep the insects off. Flies carry filth, and yellow jackets will bite off little pieces of fruit to carry away.

If the weather is too cool and damp for sun-drying, use the oven. Spread the slices out on a dry cookie sheet, stand them on edge if possible, and put them in the oven at lowest heat. Leave the door propped open just a crack to let the moisture escape. If the apple slices don't touch each other, they need no turning or stirring. It may take fifteen to twenty hours in the oven, depending on the temperature, the size of the slices, and the moisture content of the apples.

They will feel almost hard when you first take them from the cookie sheet. Put them in a plastic bag, jar, or some other closed container for storage. After a few days the moisture content will equalize, and the inner moisture will make them flexible and chewy but not soft. They will dry too much for molds to thrive and have too high a sugar level for bacteria. The concentration of natural sugars makes the apple slices as sweet as candy and much better for you. I still enjoy them raw with meals or afterward. In addition, you can stew or boil them to make excellent applesauce. For more rapid cooking, we often run the dried slices through the grinder and pack them into small plastic bags in portions just right for one meal. When camping or backpacking, it is a convenient way to carry fruit. With enough water to reconstitute them, put one package of ground dried apples in a kettle and place it on the edge of the campfire for the evening. The heat hastens the softening process, and by morning you have a kettle of delicious applesauce for breakfast.

A little ingenuity will help you dry most fruits and vegetables. If you can use the sun to do so, it may prove to be the least-expensive way to preserve food. Remember, however, that drying destroys more vitamins than either freezing or canning. It will not eliminate all vitamins but will reduce the amount per volume. As a result it may take more careful planning to ensure adequate vitamins in the diet.

Soft fruits such as apricots, peaches, and pears will dry in the same way as apples, but they may take a bit more careful handling. You should dry peach and pear halves without slicing to avoid loss of juice. Since the riper fruit has a greater concentration of sugars, it will spoil less than unripe fruit. Most fruits turn dark during the drying process and don't look appetizing, even though the flavor is excellent.

You may counteract the darkening by bleaching with sulfur fumes, which is how commercial driers produce pears and peaches and apricots that have their natural ripe color. You can do the same thing if you wish, or you can just train yourself to disregard the color and enjoy the flavor. Or you may employ antioxidants such as ascorbic acid—vitamin C—which you may purchase in a powder form to dissolve in water as a dip for the fruit just before drying. While it will not prevent darkening entirely, still it will help to retain some of the natural color. Apples dipped in ascorbic-acid solution do dry with a lighter color than those not treated.

Changes in texture and flavor are not necessarily undesirable either. Some of our friends prefer home-dried corn over either canned or frozen corn. Dried corn is easier to store than frozen corn, but it may take more care than canned corn. If you have never tried dried corn, its interesting flavor might pleasantly surprise you.

Whenever I eat dried corn, it reminds me of the old-fashioned corn roasts we used to have down by the river. We built a large bonfire on the sandy beach. When a large mass of hot coals had accumulated, we dug a pit near the fire and partly filled it with live coals. After throwing a thin layer of sand on the coals, we placed several dozen ears of corn still in the husk on it and in turn covered them with more sand. The rest of the fire we raked over the top of the pit and allowed to burn for at least an hour. If we could convince the skeptics that the corn wasn't burning up so that they wouldn't dig it out too quickly, the hot roasted ears, stripped of their husks and rubbed over a slab of butter, had the same interesting flavor as dried corn.

Corn can be sun-dried or oven-dried. Prepare it the same as for freezing or canning, that is, blanche and cut it from the cob. Then spread it out on shallow pans or cookie sheets to dry. A thin layer will dry faster with less chance of spoilage. We have dried it in the oven at lowest heat in inch-deep layers. It takes considerable stirring, however, especially at first to keep it from sticking together. Usually it takes twenty-four to thirty hours in the oven and longer in the sunshine. When sun-drying, bring it in during the night or at least cover it to prevent dew from collecting on it. Also protect it from flies and other insects with fine cloth netting or screen wire. You may want to start the drying process in the oven and complete it in the sun. When the corn rattles and has turned golden brown in color, store it in airtight containers or cloth bags. Tight containers will exclude weevils and other pests. Rehydrate it by soaking overnight in water and then cook as if it were canned or frozen. I particularly remember its distinctive flavor as scalloped corn.

Many people overlook dried beans as a convenient way to store protein. An extra row of beans in your garden can produce a good winter food supply. Let them ripen on the plants, but pick the dried pods before they split open and scatter the beans on the ground. In a dry climate the beans may be ready for storage as soon as you harvest them. If they are not fully dry, a few hours in the oven will harden them and also destroy any weevils that may have already attacked them. Kept in tight containers to protect them from later invasion by weevils, they may last for several years.

Peas, garbanzos, and lentils also survive well in the dried state, as do small grains. You will find it difficult to thresh and clean wheat, barley, and rye on a small scale. It may be best to purchase grains from the large commercial growers unless you have lots of time and ingenuity to spend on winnowing and screening. People have told me of home-built threshing and cleaning machinery, but I have not seen any such equipment listed for sale in the catalogs.

Corn is one of the easier grains to harvest and clean because it doesn't have as much contamination by kernel-sized weed seeds and pebbles as does wheat. If you allow early planted corn to ripen and harden on the ear, you can shell it off and clean it with a simple winnowing process. Fresh-ground cornmeal will give you a new taste experience in corn bread. The packaged cornmeal on the supermarket shelf has of necessity had almost all of the oil and germ removed to prolong its shelf life without rancidness. If you grind your own corn into cornmeal, you will have the whole thing—oil, germ, color, flavor, and texture. You just haven't tasted real corn bread until you make it from your own whole cornmeal. By varying what kind you grow, you can even have a wide range of flavors. My friends laughed at me when I told them I was leaving a few ears of my Golden Bantam sweet corn to ripen for cornmeal. In their minds, one could make cornmeal only from hard field corn. Since I didn't know any better, I went ahead and did it anyway. The following winter I had some of the same skeptical friends as guests at our table to share hot vegetable stew and fresh baked corn bread with honey from our bees. The full, rich flavor surprised them, and they are making plans to save some of their Golden Bantam ears for cornmeal next season.

Corn bread made from sweet corn has so much oil that you really don't need to put butter on it when you eat it warm. Don't grind up more than you can use right away unless you can store it in the refrigerator. It will turn rancid quickly if left in a warm room. I am eager now to try making cornmeal from several other varieties. The many available should keep me busy for years to come testing flavors and combinations of flavors in cornmeal.

Don't forget to include popcorn in your store of dry foods for the winter. It will keep for a long time and add a nice variety to the diet. Should it fail to pop well, try adding a little moisture. Two tablespoons of water poured over a pint of unpopped corn and kept in a closed jar will usually quickly revive the dry kernels. When heated for popping, the extra moisture will produce steam and a bigger explosion.

If you have some extra whole-kernel dried sweet corn that you don't grind into cornmeal, try parching it. Place a single layer of kernels in a frying pan with a small amount of cooking oil, and, to keep it from scorching, stir frequently while heating. As soon as the kernels puff up round and smooth, pour them into a dish and add a little salt while stirring them until they cool enough to handle without burning your fingers. They are now the crunchy kernels called corn nuts that you buy in the store. Their size, hardness, and flavor vary according to the variety of corn. Most corn nuts are made from a special variety of corn that has extra-large soft kernels. It takes pretty good teeth to eat parched corn made from the small hard kernels of most sweet corn, but the flavor is delightful. You can store parched corn and use it over a long period of time, but it seems that the flavor is best while it is still warm.

Dried prunes have acquired a strange reputation. Most people think of prunes as a medicine more than a food. True, they do have a mild laxative effect to a somewhat greater extent than most other fruits, but that is not their only virtue. They are particularly rich in vitamins and minerals, and besides that, they have a delightful flavor.

If you are fortunate enough to have a prune tree in your orchard, you can enjoy the prunes not only fresh, canned, or frozen but dry for the winter too. Let them ripen on the tree and then shake them off and gather them before the birds and insects get to them. Prunes sun-dry easily, but if the sun isn't warm enough, you may have to use the oven. When dried on a rack, the extra juice may drip out of them, but in a pan they will catch and reabsorb the juice. It will make softer fruit that is easier to chew away from the pits. Like dried apples, dried prunes have the natural sugars concentrated into a smaller package and seem sweeter than when fresh.

If you live in a climate where seedless grapes will ripen fully, you can make your own raisins. As with prunes, leave the grapes on the vine until fully ripe. Sun-drying is easiest. Thompson seedless grapes are used for most raisins, but Black Mannuka grapes turn into especially fine raisins too. I remember puffed, seeded-muscat raisins as a special treat in my childhood days, and I'm sure most varieties of grapes would convert into raisins if you work out a method for drying them. If you live in a northern climate where grapes do not ripen well, you should grow currants. Dried currants substitute for raisins in almost every way. Some people even prefer currants over raisins in cakes and cookies.

One of the oldest methods of preserving foods for winter use is making kraut. Today sauerkraut is more of a luxury than a necessity because of the abundance of food available in all seasons. But once a large crock of sauerkraut was practically the only supply of a winter vegetable one had. In some places like Korea, people ferment a mixture of vegetables rather than just cabbage.

"Krauting" is not just salt-curing. In fact, you use a limited amount of salt so that certain bacteria can multiply to produce fermentation. The acids that result not only give the sour taste to the cabbage but protect it from further decay by other organisms. But you must carry out the krauting process carefully, or you will have just so much rotten cabbage to throw away.

In most climates, well-developed, firm heads of cabbage can stay in the garden well into the winter. Even after frosts have softened the outer leaves, the tightly packed leaves of the cabbagehead will still remain crisp and sweet. We always used fresh cabbage for cole slaw and salads as long as possible and then made sauerkraut just before the hard freezing winter weather would spoil the remaining cabbage.

The practice had more than one advantage. We obtained the extra vitamins from the fresh cabbage as long as possible. Delaying the krauting meant having it at a time when we didn't have as much else to take its place. Also we would have emptied a number of canning jars by then, which we could fill with sauerkraut to get the maximum use of our food-storage facilities.

The krauting equipment that I remember was a heavy ten-gallon glazed crock with a round piece of board for a float and a large smooth stone for a weight. The kraut cutter we borrowed once a year for the brief time that we needed it. We could have sliced the cabbage with a sharp knife, but the cutter was much easier to use. It consisted of a board with a sharp blade mounted at an angle in an opening in the center. Two side rails formed a trough in which one pushed a square box holding the cabbage back and forth over the blade. Designed to cut only one-sixteenth inch at a time, it produced a uniform series of slices that would react evenly in the curing process. You can substitute a wooden barrel for the crock, but it should be all hardwood, since you don't want the resinous flavors of pine or other softwoods. Also avoid any metal or glue that the acids might affect. Clean all of the equipment carefully to avoid contamination.

Wash the cabbages and remove all damaged or decayed outer leaves. It will take about forty pounds of cabbage to fill a ten-gallon crock. Mix two cups of salt with that much cabbage as you place it gently in the crock. Don't use iodized salt, since it will destroy the fermenting bacteria. Pickling salt will be better because it does not usually contain other chemicals. Tamp the salted cabbage in firmly, layer by layer, but don't mash it.

The salt will start the flow of juices, and to keep the cabbage from floating, you must place some weight on top of it. Mother had a large plate that fit nicely into the top of the crock. A round piece of hardwood board would do as well. If you drill a few small holes partway through from the top, you can hook the head of a nail into the hole when lifting the board out later. Don't leave the nail in the board though, because it will likely corrode in the juices that may cover the board. A large clean rock (scrubbed with soap and hot water) makes a good weight to hold the shredded cabbage down in its own juice and keep out the air. A few layers of cheesecloth under the plate or board will also help to protect the cabbage from exposure to air.

In about six to eight hours after you fill the kraut barrel, the juice should have risen to the top cover. If it hasn't, add more weight until it does. It is important that the whole crock cures at the same rate. The proper temperature is between 65 degrees and 70 degrees F. If too warm, the cabbage gets slimy and spoils; if too cold, the process is so slow it may not develop good flavor. Fermentation produces bubbles of gas that you can see rising to the surface of the juice.

A white scum will form on the surface. You should remove this frequently—every day or two—by taking up the weight and board and lifting off the cheese-cloth, which will contain most of the white stuff. Rinse the cheesecloth, board, and weight in hot water and then cool before replacing them. In about four weeks, at the proper temperature, the process of fermentation will stop. No more gas bubbles or white scum will appear. The cabbage will have changed from greenish white to a translucent off-color white.

You can keep the sauerkraut in its own juices for two to three months at temperatures below 50 degrees F but not freezing. If the brine evaporates, add pure water to keep the top covered. Should you not eat it all up in a few weeks or months, you can preserve part of it for later use. Can it as you would fruit.

CHAPTER 11

TOOLS AND WORK

Whatever work you have to do on the small farm, it goes most efficiently with the proper tools. A knowledge of which one to use and how, along with how to sharpen and maintain it, is essential. What can be more discouraging than trying to grow a good garden with dull and damaged tools or with the wrong ones?

A shovel is probably the first thing that comes to mind when you think about working with the soil (unless you have the misconception that you can do nothing in the garden without a powered Rototiller or tractor). Shovels come in many shapes and sizes. The most common is a contractor's shovel for sand and gravel. It is designed for shoveling, not spading. You will find the straight-backed irrigator's shovel much easier for spading because you can hold the handle straight while pushing the blade into the soil with your foot.

Spading shovels come in various styles, mostly with a square-ended blade rather than a pointed blade. They are useful in soft soil or light sod, but if you have hard soil or heavy sod and weeds, the pointed shovel will push more easily into the ground. It is rather unlikely that you will have much use for a short-handled shovel unless you do some specialized work around shrubs, and even then a long-handled shovel will usually function almost as well. The short-handled shovel requires more stooping. A flat, square-ended scoop shovel can be handy for shoveling snow as well as picking up potting soil from a

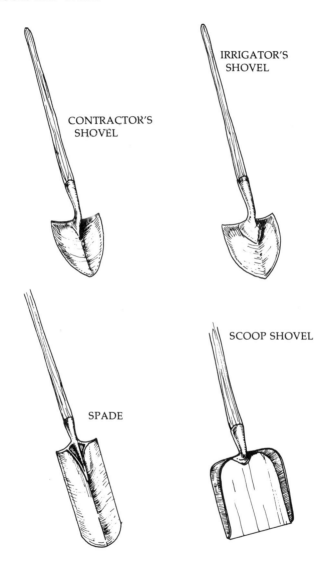

CONTRACTOR'S
SHOVEL

IRRIGATOR'S
SHOVEL

SPADE

SCOOP SHOVEL

smooth flat surface. If you want to keep your tool expenses to a minimum, a straight-backed irrigator's shovel will perform most of the jobs you will have.

Don't waste your money on a cheap shovel. It will have a weak handle that will break, or the blade will bend, and you will spend more in the long run replacing it than if you buy a good one in the first place. A shovel blade made of good steel sharpens and cleans better than one of poor quality. You can sharpen a shovel with a flat mill file. If you don't wait too long to

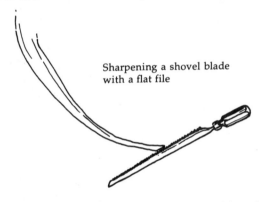

Sharpening a shovel blade
with a flat file

sharpen it, you can recognize the original bevel and remove only enough metal to repair the smooth cutting edge. Don't make the cutting edge too thin, or every little stone it strikes will damage it. On the other hand, if it is too blunt, it will not cut through sod and roots as easily. Keep it sharper for spading than for shoveling sand.

Never leave a shovel out in the weather. Rain will leave the handle checked and roughened so that it hurts your hands to hold it. Damp soil left on the blade will produce rusty spots that prevent the dirt from gliding smoothly over the blade. You can waste a lot of time and energy trying to keep soil from sticking to a rusty blade.

Have a stiff-fiber scrub brush hanging by a faucet near the door of the tool shed and use it to clean all of the dirt from your shovel and other tools as you return them. You may have to discipline yourself to develop good habits in proper tool maintenance, but it will certainly pay off in longer life for your tools and longer life for yourself, with fewer frustrations over dull, rusty, inefficient equipment. The cold rainy days of winter are a good time to work inside fixing up your tools. The shovel may need to have the handle smoothed with sandpaper or steel wool. You may wish to paint the shank with a thin coat of rust-resistant paint, but don't paint the blade. It will scratch and peel as you push it through the soil, and dirt will stick to it more readily than to the polished metal. If you live in a damp climate, store your tools in a dry place. Perhaps you may need to rub cleaned metal surfaces with a little lubricating oil for winter storage.

Include at least one hoe among your tools, even if you are a disciple of Ruth Stout. A hoe is a good cultivating tool, and most people use it to chop weeds. Don't employ a hoe designed to cut weeds at or just below the surface to dig large roots or break up heavy clods. The ordinary garden hoe consists of a blade welded to a shank fastened to the wooden handle. Its durability depends on the strength and design of the shank and how securely it is attached to the handle. Most tool sheds contain a few hoe blades that have broken from the shank or that have pulled with the shank from the handle. Such damage is difficult to repair. Try to avoid it by purchasing a sturdily built hoe and using it only for its intended task.

Hoeing requires a clean sharp edge. With a flat file you can sharpen a good hoe blade to razor sharpness. Remember, however, that you will probably strike stones and pebbles in the soil, so don't make the edge too thin. The nature of your soil will determine at

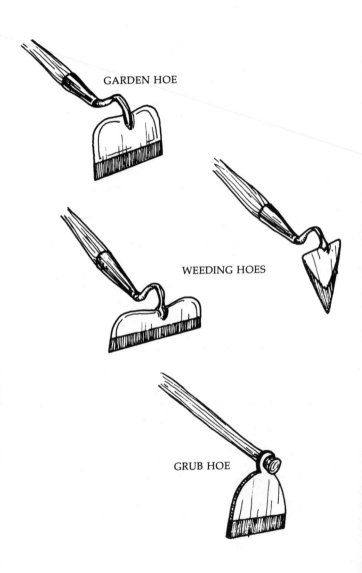

GARDEN HOE

WEEDING HOES

GRUB HOE

what angle you file your hoe blade. If it is really stony, hold the file at a wide angle with the blade so the edge will not nick and turn as easily.

Cleaning the hoe after each use is essential to keep it in good condition. The scrub brush and water will remove soil that encourages rusty spots. Keep the handle clean and dry to avoid checking and rotting. A blade sharpened so many times that it is only 2 to 3 inches long compared to its original 4 to 5 inches shows the rewards of good maintenance. Most people break the handle or the shank long before they have occasion to sharpen it that much. Some skilled gardeners have a special hoe that they like for weeding. They have it filed down to a narrow blade that can reach in between plants for hard-to-get weeds, thus saving a great deal of back bending. You can buy special weeding hoes that are narrow and pointed with a long, curved shank.

Hoes have a long handle for a good reason, and you should learn to take advantage of it. With a clean, properly sharpened hoe, you should be able to do the

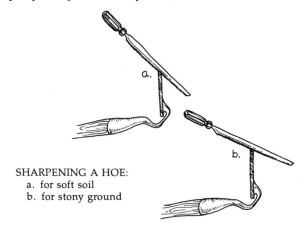

SHARPENING A HOE:
a. for soft soil
b. for stony ground

job without stooping. If you are trying to do a job too
heavy for the regular hoe, then you should switch to a
different tool. To break up heavy clods or dig out
deep-rooted sod or weeds, employ some type of grub
hoe. The name most correctly applies to a heavy steel
blade securely fastened to a thicker, stronger handle
than ordinary hoes. The blade may weigh three or
four pounds. Once you get it in motion, it will strike
with considerable force.

MATTOCK

Another heavy tool sometimes called a grub hoe,
but more accurately termed a mattock, has a heavy
grubbing blade on one side of the head and an ax
blade or pick on the other side. It is fitted to a short,
heavy handle like a pick, but the grubbing blade
makes it well suited to digging hard soil, sod, or roots.
Working with it makes a good reducing exercise. I
have found it a help in breaking up small areas of
heavy sod or loosening the packed dirt of a pathway.
It is almost indispensable in clearing land of brush
and small trees. Don't waste your time and energy
with a pick when the wider blade of a mattock will get
the job done with only one fourth as many swings of
the heavy tool. Most large hardware stores still carry
mattocks.

Rakes level loose soil or gather leaves. Most are now built for raking leaves. The lawn rake, made of split bamboo or thin metal strips, will work only for cleaning up around the yard. The longer-toothed, so-called garden rake is usually secured to the handle by means of a curved rod from the ends of the rake rather than the middle. Such a fastening allows a relatively light rake bar that doesn't bend, since the mounting pulls it from both ends, and the curved rod allows some spring so it doesn't break as readily when caught on sticks or stones. But a light rake will not pulverize large clods or pull loads of heavy soil in preparing a seedbed. Many broken and battered rakes bear testimony to that fact.

LAWN RAKE

LEAF RAKE

GARDEN RAKE

If you want a real garden rake, get one constructed of strong steel and securely fastened to the handle from the middle of the bar. The teeth will not bend when you pound clods with it, and it will bear as much of a load as you can pull. If used right, kept clean, and stored properly, it should last the rest of your lifetime.

Many overlook forks as gardening tools. They come in many shapes, sizes, and forms. Probably the most versatile for general gardening is a type of pitchfork—not the two- or three-tined hayfork but a stronger five-tined fork. The manure fork is similar but usually heavier and straighter. The slightly curved tines of a good garden pitchfork are usually about 12 inches long. The handle is at least as long as a shovel handle and as sturdy.

When you clean up the garden spot in the fall, you will have tangles of tomato and cucumber vines and other plant debris to rake into a pile. Unless you plan to make a compost heap right there, you will want to load the trash into the wheelbarrow to haul away. Have you ever tried to pick such material up with a shovel? It can be quite frustrating. A five-tined pitchfork will do the job easily and not scoop up soil and fine mulch with the trash.

When breaking up new ground, you end up with a pile of weeds and loosened sod raked to one edge. Without a lot of stooping, the fork will do the job and even allow you to shake out any good topsoil that you might have caught up with the debris. Coarse mulching material spreads easily with the five-tined fork. It also works well with partly decomposed compost that resists the shovel and wet leaves and grass that will not respond to the rake.

My father always preferred the five-tined fork for digging potatoes. Its long handle eliminated stooping. The space between the tines is small enough to pick up the potatoes but wide enough to let the dirt

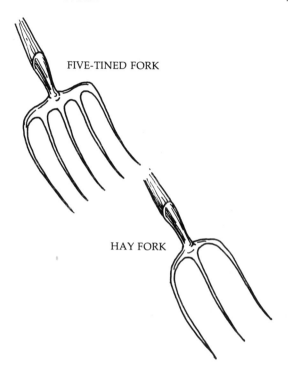

FIVE-TINED FORK

HAY FORK

drop through. The sharp tines push easily into the ground and deeply enough to catch all of the potatoes in the hill.

There is also a spading fork which has shorter, heavier tines. The tines are usually flattened and close together to pick up compact soil. It will not work with really light and fluffy dirt, but such ground probably doesn't need spading anyway; if it does, the straight-backed shovel will do a good job. I have a spading fork, but I rarely use it for that. It seems more valuable for jobs that require something in between

the five-tined fork and the shovel. For example, I dig
raspberry plants from the old patch to start a new one,
or carrots from soil so compacted that it might bend
the prongs of the five-tined fork. It also removes bulbs
from hard-packed earth.

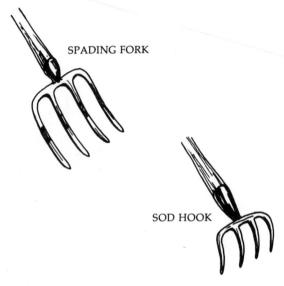

SPADING FORK

SOD HOOK

One rather common tool that looks like a cross
between a fork and a hoe is the sod hook. Usually
having four tines bent at right angles to the handle so
you can chop with it instead of lifting, it is particularly
valuable in loosening light sod or clumps of grass and
weeds. With a combined chopping and raking action,
you can shake the clumps of sod free of dirt and place
them in piles to pick up with a fork. I have found the
sod hook a good aid in breaking the soil surface compacted by heavy watering or rains. If the soil is well
tilled, the tines of the sod hook pull easily through the
surface layer, leaving it loose and crumbled.

A cover crop growing between your fruit trees or on unused garden spots may need an occasional mowing. The old-fashioned hand scythe is still practical for the job. The snath, or handle, has just the right length and angle to pull the blade against the stems close to the ground for even cutting. Such a tool mowed hundreds of acres of hay and grain before the invention of horse-drawn mowing machines and reapers. A special attachment called a cradle enabled the operator to not only cut the grain but leave it neatly arranged in bundles that others could tie into sheaves. You may not need the cradle, but the blade-and-snath unit will do a good job for you if you learn how to operate it properly. It takes a little practice, so don't give up too easily. At best it is hard work, but with experience you will learn how to swing your whole body in a rhythmical pattern that will carry the blade evenly and with sufficient force to cut a large swath. I have heard older people tell about expert scythemen who could mow a lawn and leave it as neatly manicured as any power mower made today.

You will find that tall grass and forage are easier to cut through because they don't bend over as readily when the blade strikes them. The blade must remain razor sharp, which means frequent sharpening with a Carborundum stone. The scythe stone is about 12 inches long and 1 inch wide. Carry it in your pocket, and whenever you must stop to rest, which is quite often for most people, touch up the edge of the blade with the fine stone. A dull scythe not only will cut poorly but will wear out a person much more quickly. A scythe blade for mowing hay is long and slender and relatively light in weight. You can get a blade for mowing coarse weeds that is wider and stronger but also heavier. A short blade about 18 inches long and 4 inches wide, called a brush hook, will actually handle light brush if kept real sharp.

Those small areas of grass and weeds that you find

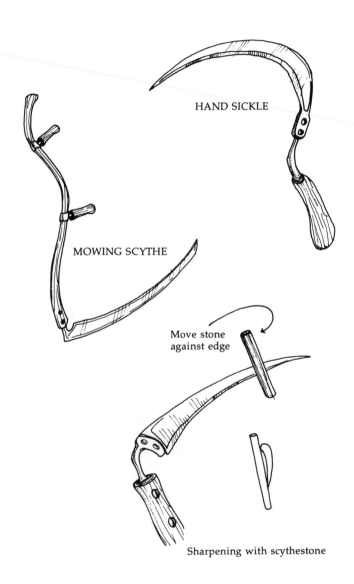

HAND SICKLE

MOWING SCYTHE

Move stone
against edge

Sharpening with scythestone

hard to get at with a scythe you can mow with a hand sickle. The sickle has a curved blade and a sharp cutting edge that you renew with a scythe stone. The sickle requires more wrist action than shoulder movement, and many short strokes are better than long pulls.

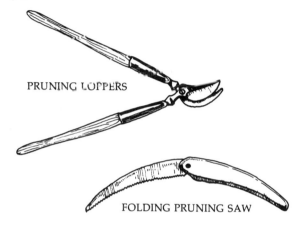

PRUNING LOPPERS

FOLDING PRUNING SAW

Every small farm will need some pruning tools. Hedge shears will do for the yard work, but fruit trees require at least a pair of hand clippers, pruning loppers, and a pruning saw. Hand clippers should be large enough to cut twigs up to three eighths of an inch and leave a smooth end. Here again, a little extra money for good quality will be an excellent investment. Pruning loppers have wooden handles 1½ to 3 feet in length that provide greater leverage for snipping off branches up to 2 inches in diameter. A folding pruning saw with a wood handle that flips over the blade to protect the teeth, and your pocket, while you carry it will handle limbs up to 6 inches in diameter. For anything larger than that you will need a regular crosscut saw or perhaps even a chain saw.

Powered equipment such as Rototillers, mowers, and chain saws require a whole array of wrenches, screwdrivers, and pliers for maintenance and repair. No small farm can function properly without a shop to store tools and to service equipment. The shop should have a sturdy workbench with a vise and clamps. The variety of jobs attempted and the funds available for tools will determine the nature of your collection. It is not difficult to get carried away with buying new and different tools for every specific purpose. Most tool collections grow over a period of years.

Plumbing, electrical work, and woodworking all require special equipment and skills to go with them. If you plan to be fully self-sufficient, you must learn the fundamental principles of many trades. On the other hand, if you can become expert in one or two, you can probably trade work with neighbors who have abilities different from yours. The latter is a common pattern in older farm communities. When you are a newcomer, you must first demonstrate some skills on your own land before you expect a neighbor to ask you to help on his place. Most public school systems offer classes that can teach you some of the basic training that you lack. Jobs of an apprenticeship type where you learn by actual experience under supervision are excellent ways to increase your proficiency in certain trades. Since few people become experts in more than a few lines, you should select one or two that fit best with your interests and abilities and pursue those until you really have something to offer in the way of work expertise. Such a skill will often turn out to be the best solution to the problem of cash income for a full-time homesteader. Building a reputation as an expert will put your services in demand to the extent that you will have to decide whether to become a tradesman and do your farming on the side or accept only enough work to provide the cash you need for taxes and other yearly expenses.

CHAPTER 12

LEISURE TIME

A common misconception about country living is that you have no leisure time. Some think that it is all work and no fun. On the contrary, a slower pace of living and a less complex life can give you a bonus of extra hours for other activities. The amount of spare time you have will depend to a large extent on how you plan your daily activities and what goals you set for yourself.

Some people seem to think they must accomplish in a few years what their grandparents spent half a lifetime doing, namely, carving a comfortable home out of the wilderness. No matter whether the wilderness is a remote mountain valley or a rural half acre, I see no sense in turning a potentially satisfying and rewarding experience into a period of nerve-shattering frustrations. Why would you choose to live in the country? Most likely because you crave the peace and quiet of the "good life" close to the soil. It is entirely possible that you could spend half of your lifetime trying so hard to make yourself "comfortable" on the land that you never have time to enjoy the peace and quiet that you have around you. Ambition is a laudable virtue if managed with temperance. The most enjoyable leisure hours are the well-earned ones. But if you find yourself too worn out to enjoy them, what have you gained?

Relaxation does not necessarily refer to hours spent rocking in the lazy-boy chair or swinging in the porch hammock. People who enjoy the challenge of

country living will find the most pleasure in projects
that are creative and result in some tangible accomplishment. Is time spent in making some useful object
really leisure time? Your answer to that question depends on attitude and priority. Cutting wood isn't a
leisure activity if it is the only way to provide warmth.
It is an essential chore. If, on the other hand, you heat
the house with oil or gas and have no real need for
firewood, but you enjoy chopping and splitting wood
when you have nothing else to do, then it could be
relaxing. Or, if you like to grow flowers and do it only
for fun, you might spend leisure hours in hard labor
developing your flower garden. Even if raising flowers is your occupation, you might enjoy some hours
working with a special kind that you do not take time
for generally.

It is often difficult to separate occupation and leisure completely. In a country-living situation, your
hobbies and special interests may merge with your
work. You don't need to apologize for enjoying your
livelihood so much that it seems like fun. The greatest
satisfaction can come from working hard at the things
you like to do. If you find a life-style in which work
becomes play and you have no drudgery, then you are
indeed fortunate. But that will not likely happen for
many of us. There always seems to be some task we
would like to escape, so we frequently reward ourselves for accomplishing the unpleasant ones by in-
dulging in something more enjoyable. When the rec-
reation results in some useful item or information that
makes our lives more meaningful, then we have a
double benefit.

Every country setting has a potential stage for observing wildlife. It doesn't have to be bear or deer or
moose to be exciting. Have you ever stepped into the
world of small creatures living in the corner of your
garden? You don't have to be a biologist in order to
learn about nature. All you need is a developing in-

terest and a willingness to sharpen your senses. Rather than watching a Disney film on TV, try going to the spot where the film might have been made and seeing it firsthand. Always you will spot something new.

I have sat on the same big rock under the apple tree on numerous occasions and witnessed a different drama each time. Each observation added to my understanding of the plot and scheme. One day I watched a dusky-winged wasp flitting from leaf to leaf as if in frantic search of something. At the end of several minutes it disappeared. Some time later I rested on the rock again, and the little wasp (apparently the same one) once more probed through the leaves as if expecting some encounter. Then I saw it pounce on a green caterpillar feeding on the leaves. I leaned closer. The little wasp expertly used its sharp sting to paralyze the caterpillar. After some apparent ceremonious dancing, it took up the prey in its jaws and like a puppy with a shoe began to carry the caterpillar away.

It could not fly with its burden, and the tortuous journey up one stem and down another kept me fascinated. At first it started out in a definite direction, but all of the ups and downs appeared to have changed its course. Suddenly it dropped the green caterpillar and flew rapidly to a bare spot of ground about 6 feet away. A few circles of orientation flight brought it to a tiny opening in the dirt. Nervously it entered, fluttering its wings. A few moments inside seemed to reassure it that it had found the right place. Then it flew directly back to the paralyzed caterpillar and started tugging it toward the hole. The wasp had to make two more orientation flights over the tangled vegetation before it achieved its goal and laid the prey in front of the opening in the ground. After scratching out a few loads of loose dirt, it dragged the caterpillar in, no doubt as a food supply for a new generation of wasps.

In those few minutes of observation I gained some insight into nature's way and made the acquaintance of an insect friend that helps me in the never-ending battle of protecting my crops from hungry pests. What would be your first reaction to see a wasp crawling at your feet? Stomp on it before it does some damage! Unfortunately that is the normal response among those who haven't spent some leisure time observing the wildlife in the corner of their garden.

Some of the country-living skills that are no longer essential to our present economy can become satisfying leisure-time activities. For example, we no longer have to spin our own wool into yarn and fabricate it into garments. Yet spinning and weaving can be most satisfying leisure activities. You can purchase numerous instruction books now, but you would do best to find someone who could teach you the basic skills. Many communities have classes in spinning and weaving, and the colleges and universities are adding such courses to their offerings.

Should you have already started to learn how to spin, you will look for usefulness rather than antique beauty when you buy a spinning wheel. Several years ago my wife learned about a company in New Zealand which makes an inexpensive, utility-model spinning wheel that, including shipping costs, was less than the ornate ones usually found for sale. It came as a kit but required only simple hand tools to assemble, and it runs beautifully. With increased interest in spinning as a hobby, more such models should become available.

If you have room to graze a few sheep, you can produce your own fleece. But unless you have land that you don't need for anything else, you could probably purchase the raw wool at less cost than raising sheep. Some people like a few sheep around as pets. Little lambs are nice, but I have never felt any fondness for grown sheep. Don't let anyone talk you into having them until you have thoroughly investigated the cost of feed and shelter, plus such problems as disease and special care. Shearing sheep is a special skill that you can learn, but it requires practice for any degree of proficiency. Some breeds of sheep produce short wool that is difficult to spin. The climate in which you live can affect its quality also.

After you have solved the problem of obtaining fleece one way or another, you will then need to prepare it for spinning. Carefully spread it out on a flat surface so you can pick out burs and other foreign matter. Remove any tag ends that are too short or extremely stained and dirty. Then sort the remainder of the fleece according to length and quality. Place part of the fleece in a cheesecloth bag and wash it gently in lukewarm water, preferably without soap so as not to remove the lanolin. Lanolin is the natural oil in the wool, and it is much easier to spin with it present.

You must get the dirt out before spinning, though, and if the fleece is extremely dirty, you may have to apply a little soap in the first washing. Dry the fleece completely, then gently spread it out on a screen. Avoid matting the fleece. Putting it out in the fresh air and sunshine is the best drying method. Next comb—or card, as it is correctly termed—the fleece. The carding equipment consists of two paddle-shaped boards covered with stiff metal bristles. Put a little of the fleece on the cards. By drawing them against each other, separate and straighten the wool fibers so they all go the same direction. Then roll them into a loose structure called a rolag.

The spinning process involves pulling or drawing out from the rolag a continuous strand of wool fibers of the desired size and twisting them together like a cable for greater strength. One way of doing that is with a drop spindle. It is merely a piece of wood that acts as a weight on the end of the thread pulled from the rolag. You twirl it like a top to twist the forming thread. Sounds easy, doesn't it? It looks simple too as you watch deft fingers trained to draw out the fleece and simultaneously place just the right tension and motion necessary to make a smooth, even thread. Someone has said that if you can pat your head and rub your stomach at the same time, you can do well at spinning.

SPINNING WHEEL

The spinning wheel is a refinement of the drop spindle that allows you to keep the "stick" spinning by using your feet. The treadle turns the large wheel, and a belt connects it to the spindle. Each turn of the large wheel makes the spindle revolve many times, pulling the thread from the rolag onto the bobbin of the spindle as it turns rapidly and giving the twist to the thread or yarn. Operating a spinning wheel is a skill that requires considerable practice, so don't expect to start right out producing beautiful skeins of plied yarn that you can knit into fancy sweaters. We are talking about a leisure activity, so do it for enjoyment. Probably the greatest pleasure will come when you can show your friends a lovely shawl or garment that you have had the satisfaction of producing from raw fleece.

It will take more than just spinning, though, to turn fleece into something useful. After you have spun the yarn, you must wash it with soap to remove the lanolin and perhaps dye it to give it a color other than that of the sheep that grew it. Working with natural dyes is another fascinating hobby. Here is your chance to express originality and creativity. In the common things around you can be found many beautiful colors just waiting for you to blend into the yarn in pleasing combinations. Plant dyes are probably the easiest to obtain. Leaves of the trees are mostly green because of the large amount of chlorophyll in them. Different kinds of trees have various combinations of other pigments besides chlorophyll, however, and these pigments can give you colors for the yarn. For instance, elm leaves can produce a lovely yellow, while leaves from the purple plum contain an avocado green. Other parts of plants possess additional colors, like brown from walnut hulls or bark from alder trees, blue from indigo leaves, and red from madder roots. Combinations of yellow and blue can result in true greens. Ripe black cherries give you purple, and carrot tops offer a bright true yellow. Most such colors will not take to the wool unless you first treat the fibers with a mordant such as alum, chrome, tin, copper, or iron. Because the colors will vary considerably with each different mordant, you can see the almost endless possibilities. You make the dyebath by cutting up and boiling the plant materials and then straining out the remains to get a clean, colored fluid. Simmer the wool yarn at about 180 degrees F in the dyebath and then dry. Be careful about rapid temperature changes, as it may cause matting. Sunlight will fade some colors but not affect many others.

With the colored yarns you may knit socks or crochet afghans. The next step would be to weave your own fabric on a loom. There are little looms and big ones, simple looms and complicated ones. Before

you go any further, you may need to reach a decision
as to whether you want to keep it a hobby or turn it
into a moneymaking project.

People often consider growing flowers as a
leisure-time activity because of their nonutilitarian
nature. Flowers may not add much in the way of food
and other essentials, but they give a special dimension to country living that often relieves drudgery and
makes home a pleasant place. The early settlers who
pioneered the difficult country knew the value of
flowers. Often the site of an old homestead place has
as its memorial a persistent climbing rose or patch of
hardy foxgloves. People lavished painstaking care on
a few cuttings of favorite houseplants as they moved
from one frontier home to another, sometimes over
long distances. We have a Christmas cactus that is a
direct descendant of cuttings from a plant that my
wife's grandmother lovingly protected through many
North Dakota winters.

Hobby plants such as African violets, ferns, and
orchids can bring great satisfaction to "specialists"
who like to excel in something. After you become
proficient in bringing them into bloom, the next step
is usually collecting new varieties, then beyond
that—if you want to do so—developing your own new
ones. If you desire, your leisure-time pursuit may
turn into a moneymaking scheme when you begin
propagating and selling. It can still remain a relaxing
activity, however, if you don't allow it to grow into a
regular business.

Making things out of wood, from wood carving to
cabinetmaking, is a favorite during the winter when
weather prevents the usual outdoor work. It doesn't
take a lot of expensive equipment to start a woodworking hobby. A few simple hand tools can provide
the essentials for many pleasant hours. However, like
anything else, as you become more adept, you will
want more and better tools. Skill in working with

wood can become a real asset to the homesteader. Aside from leisure enjoyment, it can make the difference between a well-kept home and a poorly maintained one.

The country way of life is not antisocial. Rural communities provide opportunities for cultural development and social activities. Parties and get-togethers, whether for entertainment or improvement, constitute an integral part of rural living. Long winter evenings offer leisure time for visiting and entertaining guests. Sharing experiences and ideas helps bind the community together. Summer outings and picnics provide wholesome activities for young and old. Those who move from the city to the country need not feel like they are leaving the life of culture behind. Music and the arts are part of the country way also. No home in the country should be without musical instruments. I have listened to some of the best in classical music performed by talented country people. Usually a higher percentage of the rural population enjoy and practice the fine arts than do those in the large cities.

Working with the soil does not really make people coarse and rough. Actually one can find more opportunity for refinement in the country way than in the city way. Individuals who neglect to develop character in the country would probably have done worse in the city. The beauty and peace of the countryside engenders gentleness and quietness in those who allow its influence to permeate their day-to-day activities. Unfortunately, some always search for the crude and unlovely side of life. They will find it no matter where they go.

CHAPTER 13

MOVING OUT

A recent agricultural publication entitled *Is a Family Farm the Answer?* gives the following counsel:

"Will it cut down on food costs? Perhaps . . . IF someone has the time, tools, inclination, and ability to plant and tend a family-size vegetable garden; harvest the produce; can or freeze enough of it to last through the winter. But it's debatable (all things considered) whether this is less expensive than buying the material as needed from the local market.

"Pretty much the same applies to trying to raise your own fruit, meat, eggs, or dairy products. Can you get in on the receiving end of the high food prices? Not necessarily. One of the costly lessons you may learn is that higher food prices at retail do not always mean higher returns for all farmers.

"Can you get away from the city's 'rat race' and be your own boss? Definitely yes . . . but you may find that you traded an urban rat race for a rural treadmill that leaves scant time for anything but hard work. You may find that instead of owning a farm, the farm owns you.

"So at this point, it is difficult to resist a few philosophical observations: Anyone who has or can raise the capital it takes to buy raw land and develop it into a farm probably shouldn't have to worry too much about the price tags at the food market.

"Anyone taking over an existing small farm would do well to delve deeply into the reasons why the former owner wanted to sell it" (Edward A. Yeary, *Is a*

Family Farm the Answer? University of California, California Agricultural Experiment Station, 1973).

Every individual must reach his own decision as to whether or not to move out of the city only after careful thought. A hasty one can lead to frustrating experiences and possibly even disasters. City people contemplating country life tend to be idealistic and spend much more time dreaming of the benefits than of the problems. Too often, those who are the most anxious to move out are the least prepared both in financial stability and practical experience.

One of the first considerations is finances. No one should beguile himself into thinking that he can manage a simple country life-style with all of its self-sufficiency on no cash income. No matter how much of modern technology you reject, you still need money for taxes, medical expenses, transportation, and basic equipment. The most feasible approach at the moment seems to be to select a homesite in the country without at the same time giving up a regular job for cash income. While it may mean only suburban living for many people, one should not overlook the possibilities of a country way of life on one acre or less. Those not well trained in farming would do well to practice on a half acre until they develop the skills and techniques of proper soil management and cultural practices.

Some daring and dedicated individuals have made the leap from the inner city to the remote countryside with success. But most likely they have had some previous experience at country living or were willing to exist at poverty levels until they established themselves. Of course, a fixed income such as retirement benefits can make up the difference between survival and comfort for many people. It doesn't seem advisable, however, to wait until retirement to move out of the cities. In fact, the majority of retired people need the security and advantage of close neighbors and the technical assistance available primarily in a nonrural community. Besides that, it is a well-accepted fact that the rural environment is the best for raising children. So, it is the younger generation that particularly should find ways to leave the cities. We should add, however, that the most contented elderly people are those that learned the country way of life in their younger years and now enjoy good health and independence on a small piece of land.

Those who have a real desire to exit the cities would do well to learn as much as possible from the experiences of those who have tried it, both successfully and unsuccessfully. Your country-dwelling friends can be your best teachers. You will have to adapt their techniques to fit your own life-style, but their successes and failures can supply you with both courage and caution.

How self-sufficient can you really be in a truly rural environment? That question involves several others, such as what level of income will it take to satisfy your needs, and how many of the so-called advantages of technology are you willing to give up? Could you be contented spending your entire time working with the soil to produce food, or do you have a professional or trade skill that brings you greater satisfaction?

Some people have demonstrated complete self-sufficiency on a small piece of land, but their life-style may not satisfy you. Therefore, you must be willing to change yours or find a way to both live on the land and retain your current one. Some have become so disenchanted with the artificial, superficial, technological approach to life that they want to switch over to a more natural, simple way. But that is not as easy as many think, and it will require large amounts of faith and persistence.

A cash income allows you more choice in life-style.

You may elect to spend your money on fancy automobiles or laborsaving gadgets, or you may decide to invest it in country property. If you can live in the country and still retain your cash income, you have the opportunity to explore the country way just as far as you wish. But what about those who must give it up in order to move to the country? Can they make it on the land? Some have done it, but can you?

Be careful about the glowing accounts of country living that you find in books and magazines. An individual who can successfully write a book about his country-living experience and have it published is probably a professional writer and has considerable cash income. His rural sojourn may have been only long enough to give him material for a good book. What about the people who have left their city jobs and moved to the country to make a living on the land? If they have enough capital to purchase their own land and be free of debt, they don't require a large cash income. Also they may be able to raise a salable crop on their land or find a part-time job working for someone else.

A crop that earns enough profit to support a family for a year requires skill, experience, and an initial investment in land and equipment. Most farmers grow such crops in a monoculture system on a large scale to take advantage of efficient machinery. It reduces labor costs and enables them to sell large volumes at small margins and still make a living. If you want to compete with agribusiness, you must also become a large producer, or you must have a salable item that agriculture cannot easily raise in large mass-produced quantities. What are such items?

Can you think of a crop that machinery cannot plant and harvest? If so, you may be on the right track. What about strawberries and raspberries? They might be a good cash crop if you have skill at growing them and can find a market. Cherries could fit in such a category also, but it takes many years to get trees large enough to bear much. You could temporarily raise small fruits such as berries and grapes between the trees until the cherries come into production. You may have a hard time finding a market for your cash crop, unless you live within easy driving distance to a large city.

Roadside stands are a common method of selling home-grown produce. Gas shortages may reduce sales, however, and competition with neighbors may lower prices, but it still may be the best way to market your product. A business small enough to be a family project will free you from having to hire help, and you can retain all of the profits within the family. The amount of cash income from such a venture will be relatively small, but it could be all that you need if you are really working toward self-sufficiency.

Many people now recognize the superior quality and flavor of fresh fruits and vegetables purchased directly from the small grower and may even pay a higher price at the roadside stand than at the supermarket. Some produce such as corn and peas that have their best flavor when fresh will attract customers who may also purchase other items at the same time. In addition, you may discover certain varieties of fruits and vegetables that are fragile and difficult to ship even short distances when ripe. If you grow the varieties that do well in your climate and have a superior flavor, you may be able to sell all you can raise.

Like any other business, success depends upon ingenuity, skill, and honesty. Learn all you can from books and periodicals and take advantage of the advice offered by your county agricultural extension service. People need the services of those with experience, and you may find numerous ways to make money once you establish a reputation of "knowing how" to prune fruit trees, set out shrubs, repair equipment, or do simple construction work.

Country living does not have to be a dead-end street to poverty, but neither does it guarantee an easy living. Just as much success and failure exist among country people as city dwellers. Those with an ambition to develop talents will find opportunities wherever they go. And those who think the world owes them a living will starve just as readily anywhere.

The people who seem to get the most satisfaction out of country living today are the back-to-nature or the organic-living enthusiasts. Could it be because they recognize more than just the dollar and cents values of living on the land? I know some extremely discontented farmers who have the latest in automated equipment and laborsaving devices which they use to grow a guaranteed crop each year. But they spend more time worrying about prices and taxes than they do contemplating the benefits of the country way of life.

The organic gardener will more likely raise things for pleasure rather than for profit. He is more inclined to experiment with new ideas (or old ones) and does not concern himself so much about the cost of labor and the margin of profit. The modern, successful farmer expects and demands the same level of income as his counterpart in other industries. He is so concerned about "making it" that he usually overlooks the more intangible values such as independence, clear air, and the peaceful countryside. The organic gardener is not necessarily more capable of enjoying the smell of freshly turned earth, apple blossoms in May, or the sight of maturing crops. It's just that he takes more time for it. He is living close to the earth for pleasure more than for profit.

Is it possible to have your cake and eat it too? Can a person make a living while enjoying the full benefits of country life? I know quite contented farmers who work hard at making a living and are successful financially to the extent of owning a comfortable

home, driving a good car, and having modern
equipment. What makes the difference between satisfaction and discontent? It seems as though attitude
must have a lot to do with it. Appreciation of rural
living does not depend then so much on what you
have as on how you think. It is a way of life, to be lived
by the backyard gardener or the corporate farmer. If
your ambition is to make a livelihood by tilling the
soil, then prepare yourself as you would for any other
occupation and you may succeed as well. But if you
want the *pleasures* and *satisfactions* that country living
can bring, then study and learn as much as you can
about the country way. You will have no limit to the
enjoyment you can receive.

Some people read with fascination the delightful
experiences of others in the country, but upon arriv-
ing there they experience only dusty smells, obnox-
ious insects, and burning sun. It is true that rural life
consists of many little experiences both pleasant and
unpleasant, but the great satisfaction comes from the
complete pattern of living close to the earth, understanding the basic processes of life, and having a part
in producing the necessities of life such as food and
fiber and even life itself.

If the simple, basic aspects of the country inspire
you, you may be ready to move. It may involve leaving the city or possibly only entering a new pattern of
thinking and doing. In any case, be ready to open
your mind and your senses to the world of nature or
you will miss much. It is by reading the Bible and
contemplating the works of the Creator that we learn
more about Him. In learning more about Him we
reach a better understanding of what life is all about.
As we come to our senses, we begin to see the reflec-
tion of the love and kindness of the Creator in His
work. That develops a desire in us to be more kind
and loving. Only by cooperating with the Creator can
we develop such virtues. As we discover how to

orient our lives according to His plan, we will also know how better to care for the soil and maintain healthy gardens, healthy bodies, and healthy minds. We have a long way to go, but He is eager to lead us to a better way. Here is what country living is really all about.

It is so much easier to communicate with the Creator in the garden than on a busy street. Remember, He placed Adam and Eve in a garden and gave them work to do in developing a home in the country. He planned to teach them how to care for the soil and how to grow their own food. They were to develop character in themselves and their children that would enable them to live in perfect harmony with each other and all of nature. While we have departed a long way from that plan, the Creator is still ready to help us return, if we will follow His instructions. How rapidly we regain that original plan depends to a large extent on our faith and our willingness to respond. Great things await those who place themselves in a right relationship with the Creator and study the lessons of the plants and animals and the soil. The things they learn will bring them peace, satisfaction, and contentment in the country way of life.

BIBLIOGRAPHY

Angier, Bradford, *Free for the Eating.* Harrisburg, Pennsylvania. Stackpole Books, 1966.

Bickford, E. D. and S. Dunn, *Lighting for Plant Growth.* Kent State University, 1972.

Cohan, Ray, *How to Make It on the Land.* New Jersey. Prentice Hall, 1972.

Coleman, Peter, *Wood Stove Know How.* Charlotte, Vermont. Garden Way Publishing Co., 1974.

Colinvaux, Paul, *Introduction to Ecology.* New York. John Wiley and Sons, Inc., 1973.

Dadant, C. T., *First Lessons in Beekeeping.* American Bee Journal, 1967.

Eaton, Jerome A., *Gardening Under Glass.* New York. Macmillan, 1973.

Gay, Larry, *The Complete Book of Heating With Wood.* Charlotte, Vermont. Garden Way Publishing Co., 1974.

Gearing, Catherine, *Field Guide to Wilderness Living.* Nashville, Tennessee. Southern Publishing Association, 1973.

Kains, M. G., *Five Acres and Independence.* New York. New American Library, 1973.

Klein, G. T., *Starting Right With Poultry.* New York. Macmillan, 1947.

Langer, R. W., *Grow It!* Saturday Review Press, 1972.

Logsdon, Gene, *Homesteading: How to Find New Independence on the Land.* Emmaus, Pennsylvania. Organic Gardening and Farming, 1973.

Loveday, Evelyn, *The Complete Book of Home Storage of Vegetables and Fruits.* Charlotte, Vermont. Garden Way Publishing Co., 1972.

MacManiman, Gen, *Dry It—You'll Like It.* Fall City, Washington. Living Foods Dehydrators, 1973.

Moral, Herbert, *Buying Country Property.* Charlotte, Vermont. Garden Way Publishing Co., 1972.

Nearing, Helen and Scott, *Living the Good Life*. Schocken Books, Inc., 1970.

Ogden, Samuel, *This Country Life*. Emmaus, Pennsylvania. Rodale Press, Inc., 1973.

Ortloff, H. Stuart, and Henry G. Raymore, *A Book About Soils for the Home Gardener*. New York. William Morrow, 1972.

Robinson, David E., *The Complete Homesteading Book*. Charlotte, Vermont. Garden Way Publishing Co., 1974.

Root, A. J., *Starting Right With Bees*. Medina, Ohio. A. J. Root Co., 1971.

Rutstrum, Calvin, *The Wilderness Cabin*. London. Collier Macmillan, 1961.

Shoemaker, James, *Vegetable Growing*. New York. Wiley, 1953.

_____, *Small Fruit Culture*. New York. McGraw Hill, 1955.

Sunset Books, *Basic Gardening Illustrated*. Menlo Park, California. Lane Books, 1972.

_____, *Vegetable Gardening*. Menlo Park, California. Lane Books, 1972.

_____, *Sunset Guide to Organic Gardening*. Menlo Park, California. Lane Books, 1973.

Swan, Lester A., *Beneficial Insects*. New York. Harper and Row, 1964.

U. S. Department of Agriculture, Agriculture Yearbook. *Insects*. Washington, D.C. U.S. Government Printing Office, 1952.

_____, Agriculture Yearbook. *Plant Disease*. Washington, D.C. U.S. Government Printing Office, 1953.

_____, Agriculture Yearbook. *Seeds*. Washington, D.C. U.S. Government Printing Office, 1961.

_____, Agriculture Yearbook. *A Place to Live*. Washington, D.C. U.S. Government Printing Office, 1963.

_____, Leaflet 559, *Firewood for Your Fireplace*. Washington, D.C. U.S. Government Printing Office, 1974.

White, Ellen G., *Education*. Mountain View, California. Pacific Press Publishing Association, 1942.

_____, *The Ministry of Healing*. Mountain View, California. Pacific Press Publishing Association, 1942.

_____, *Country Living*. Washington, D.C. Review and Herald Publishing Association, 1946.

Wigginton, Eliot, *The Foxfire Book*. New York. Doubleday and Company, Inc., 1972.

_____, *The Foxfire Book II*. New York. Doubleday and Company, Inc., 1973.

Seed and Nursery Catalogs

Stark Brothers Nurseries. Louisiana, Missouri 63353. Fruit and nut trees.

J. E. Miller Nurseries, Inc. Canandaigua, New York 14424. Fruit and nut trees.

Gurney Seed and Nursery Company. Yankton, South Dakota 57078. Vegetable seeds, plants and trees, equipment, flower seeds.

W. Atlee Burpee Company. Philadelphia, Pennsylvania 19132; Clinton, Iowa 52732; Riverside, California 92502. Vegetable seeds, plants, equipment, flower seeds.

George W. Park Seed Company, Inc. Greenwood, South Carolina 29647. Flower and vegetable seeds, plants, equipment.